Sitosterol

Monographs on Atherosclerosis

Vol. 10

Editors
T. B. Clarkson, Winston-Salem, N.C.
David Kritchevsky, Philadelphia, Pa.
O. J. Pollak, Dover, Del.

S. Karger · Basel · München · Paris · London · New York · Sydney

Sitosterol

O. J. Pollak, Dover, Del.
David Kritchevsky, Philadelphia, Pa.

7 figures and 36 tables, 1981

S. Karger · Basel · München · Paris · London · New York · Sydney

Monographs on Atherosclerosis

Vol. 6: *William T. Beher* (Detroit, Mich.): Bile Acids. Chemistry and Physiology of Bile Acids and their Influence on Atherosclerosis. XIV+226 p., 11 fig., 9 tab., 1976. ISBN 3–8055–2242–8

Vol. 7: *G. A. Gresham* (Cambridge): Primate Atherosclerosis. VIII+102 p., 6 fig., 1976. ISBN 3–8055–2270–3

Vol. 8: *H. Engelberg* (Beverly Hills Calif.): Heparin
R. W. Robinson; I. N. Likar, and *L. J. Likar* (Worcester, Mass.): Arterial Mast Cells VIII+116 p., 2 fig., 6 tab., 1978. ISBN 3–8055–2892–2

Vol. 9: *H. S. Sodhi; B. J. Kudchodkar, and D. T. Mason* (Davis and Sacramento, Calif.): Clinical Methods in Study of Cholesterol Metabolism. XII+167 p., 17 fig., 2 tab., 1979. ISBN 3–8055–2986–4

National Library of Medicine, Cataloging in Publication
Pollak, Otakar Jaroslav
Sitosterol
O. J. Pollak, David Kritchevsky. – Basel, New York, Karger, 1981.
(Monographs on atherosclerosis; v. 10)
1. Sitosterols   I. Kritchevsky, David, 1920–   II. Title   III. Series
W1 MO569T v. 10/QU 95 P771s
ISBN 3–8055–0568–X

Contents

Part II

Contents VII

Part IV

Introduction

Introductions are written to justify publications. Often, the introduction takes on the character of an apology, especially if the topic is controversial.

Repetition here of a few worn phrases is unavoidable: 'For the individual, an elevated blood cholesterol *per se* is neither diagnostic nor prognostic of atherosclerosis; nor does a normal cholesterol level have reciprocal implications. However, epidemiologic data tell us that the incidence of atherosclerosis and its main sequel, myocardial infarction, is four times more frequent in persons with an elevated plasma cholesterol than in persons with normocholesterolemia. The presence of cholesterol in atherosclerotic lesions cannot be disputed. Cholesterol synthesis is virtually absent in the arterial tissue; hence, it cannot account for the considerable amounts of cholesterol in lesions. Blood has to be considered as its major source.'

In 1958, *Pollak* [931] questioned the necessity of correcting hypercholesterolemia in every instance: 'Each individual seems to have its own "basal" plasma cholesterol level below which it cannot be depressed without triggering homeostatic cholesterol synthesis and also a "maximal" level which can hardly be exceeded by an overload of dietary cholesterol.' One of the keenest students of atherosclerosis, the late *Russell L. Holman,* reacted to this statement with the following questions: 'Will you agree that the problem of atherosclerosis would be non-existent or, at least, minimized if there would be no cholesterol in the arterial lesions? If you agree, does it not follow that we must try to reduce the amount of cholesterol available for deposit in the arterial wall?' In this context it does not matter whether one considers cholesterol to be the only villain, the chief villain, or one of several, nor whether it is the primary cause or a secondary factor in atherogenesis and in the evolution of the atherosclerotic plaque.

In 1951, *Barr et al.* [52] pointed out that the ratio of α/β-lipoprotein (α/βLP) cholesterol might be more indicative of risk than the total serum cholesterol. This observation has been reinvestigated and, in 1975, *Miller and Miller* [591] elaborated on the new trend of examining the amount of cholesterol present as high-density LP (HDL) cholesterol and of judging risk accordingly.

Successful experimental induction of atherogenesis via hypercholesterolemia – to nearly complete exclusion of other etiologic factors – is considered by many as proof of a causal relationship, and hence as a justification of efforts to arrest the progression of anatomical and clinical atherosclerosis by reducing the plasma cholesterol level.

One can argue that we should treat patients rather than their symptoms, such as hyperglycemia, hyperuricemia, or hypercholesterolemia. In the latter case the argument weakens if we ascribe to cholesterol an etiologic role in atherosclerosis.

The use of phytosterols, or, specifically, of β-sitosterol, to reduce hypercholesterolemia is just one of many approaches to regulate cholesterol metabolism. We intend to put the practical use of β-sitosterol as cholesterol depressant into proper perspective. To do so, we found it desirable to expand the text by including animal experiments and by discussing non-pharmaceutical approaches. Background data on cholesterol analogues in the flora and fauna form the first part of this book. The second part is devoted to the influence of plant sterols on hypercholesterolemia.

Delineation of the topic to be covered in the second part of this book is hampered by a confused and confusing terminology. At first, it seemed that discussion should be limited to β-sitosterol as it interrelates with cholesterol metabolism and, indirectly, with atherosclerosis. It soon became clear that other phytosterols, especially stigmasterol and campesterol, have to be included. Among the plant sterols, special attention has been paid to soybean sterols. Indeed, the very first report, by *Peterson* [646], in 1951, concerning the reduction of serum cholesterol in chickens, dealt with 'soybean sterols'. A year later, *Pollak* [660] presented the first report on 'sitosterol', *per se,* as an anticholesterolemic agent in rabbits and as a hypocholesterolemic agent in man.

Once we decided to include soybean sterols it became obvious that other vegetable sterols had to be included. This meant consideration of vegetable fats and oil. *Herting* [383] who published five installments of an *Annotated Bibliography on the Effect of Phytosterols on Cholesterol Metabolism and an Atherosclerosis,* between 1957 and 1963, included essays on vegetable sterols.

Next, a discussion on the mechanism by which vegetable oils influence the plasma cholesterol level had to be added, especially as one of the concepts holds β-sitosterol responsible for the hypocholesterolemic effect of vegetable oils. The bibliography on experimental and clinical use of vegetable oils and on the mode of their action is seemingly endless. No attempt is made to cover it all in this book. Only the most provocative essays will be discussed.

Part I

A. Phytosterols and Zoosterols

The division of sterols into phytosterols and zoosterols is no longer tenable. It was based on the obsolete belief that certain sterols appear exclusively in plants and others exclusively in the animal world.

Since 1957, there has been an increasing number of reports on the occurrence of cholesterol – the classical zoosterol – and of steroid hormones in plants. Since 1962, the three major phytosterols – β-sitosterol, stigmasterol and campesterol – have been identified in tissues and blood of invertebrates and vertebrates.

Moreover, phyto- and zoosterols have been found in one and the same plant, or in one and the same animal. Then, γ-sitosterol was proven to be a mixture of β-sitosterol and cholesterol. Zoosterols have been shown to contain phytosterols.

Further blurring of the dividing line between phyto- and zoosterols occurred when it was revealed that some plants are able to synthesize cholesterol from phytosterols and that many insects and even some higher species are able to metabolize phytosterols or to convert them to cholesterol.

The differences remain quantitative: the amounts of cholesterol, or of 24-hydroxycholesterol, in plants are small and the amounts of the three major phytosterols in animal tissues and blood are also small. Though conscious of the pitfalls of the old terminology, the terms phytosterols and zoosterols are nevertheless still being used for reasons of expediency.

1. Early History of Phytosterols

Cholesterol had been known for nearly half a century when, in 1862, *Beneke* [66] discovered a sterol in peas and assumed it to be cholesterol. A book on the occurrence and distribution of cholesterol in animal and plant organisms was published by *Beneke* [65] that year.

Hesse [386], in 1878, was first to coin the term 'phytosterin' for the sterol isolated from *Calabar* beans. He considered it an isomer of cholesterol. This was questioned 50 years later by *Schönheimer et al.* [744] who refused to consider the 'isosterols', i.e. sito-, stigma- and brassicasterol, as true isomers of cholesterol.

In 1889, *Tanret* [854] reported the discovery of 'ergosterin' in the ergot fungus of maize, *Claviceps purpurea* L. The name 'sitosterin' was used in 1897 by *Burián* [155] for the sterol from cereal grain ('sitos' is Greek for grain). The same year, *Thoms* [867] applied the term 'phytos-terine' (German for phytosterols) to all plant sterols. *Hesse's* [386] phy-tosterol may well have been a mixture of sitosterol and stigmasterol. 'Stigmasterin' was found by *Windaus and Hauth* [936] in 1906 in the *Calabar* bean, *Physostigma venenosum* – hence its name.

For many years all interest focused on sitosterol and stigmasterol since these were found in an ever-increasing number of plants and seeds. Their respective empirical and structural formulas were established in 1930/31 by *Sandquist and Bengtsson* [719] and *Sandquist and Gorton* [720]. At that time, identification of sterols was based on the determination of melting point and optical rotation of pure sterols and of their acetates. The data established by these authors for stigmasterol – but not for sitosterol come close to those considered valid today.

The structural and empirical formulas of many sterols are still not known, but those of a very large number of sterols have been firmly established. The methodology has changed dramatically and comprises a whole armory of sophisticated apparatus and procedures. Today's knowl-edge of sterols is based on molecular weight (MW), molecular rotation ($[M]_D$), optical rotation ($[\alpha]_D^t$), rotary dispersion (RD), nuclear magnetic resonance (NMR), melting point (mp), infrared (IR), ultraviolet (UV) and absorption spectra (AS), on mass spectrophotometry (MS) and on the results of various chromatographic methods – paper (PC), column (CC), thin-layer (TLC), gas (GC), gas-liquid (GLC), or radio gas chromatography (RGC). These methods, and the monitoring of labeled sterols from entry into the body to their elimination, have greatly enhanced our understand-ing of sterol pathways.

2. Structural and Empirical Formulas

The list of known sterols is very large and it is still growing. In the frame of this text one can make only a limited selection of sterols and steroids. The formulas of cholesterol (fig. 1) and sitosterol (fig. 2) differ only in the side chain. Many other sterols resemble cholesterol in the ring structure but differ in the side chain. The side chain structures of a number

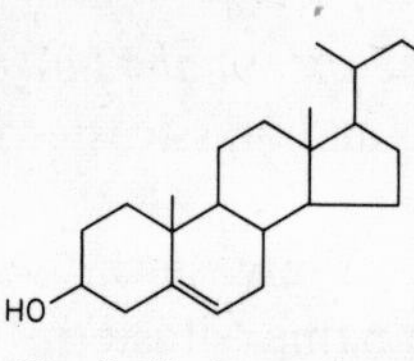

Fig. 1. Cholesterol.

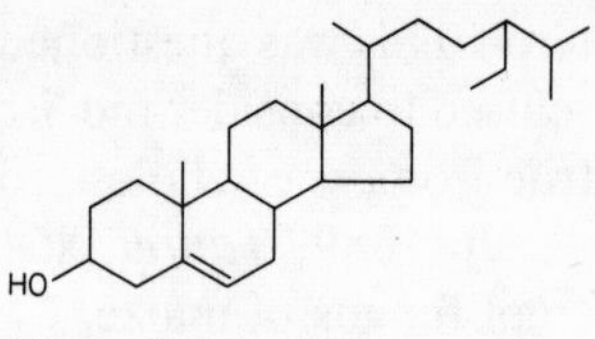

Fig. 2. Sitosterol.

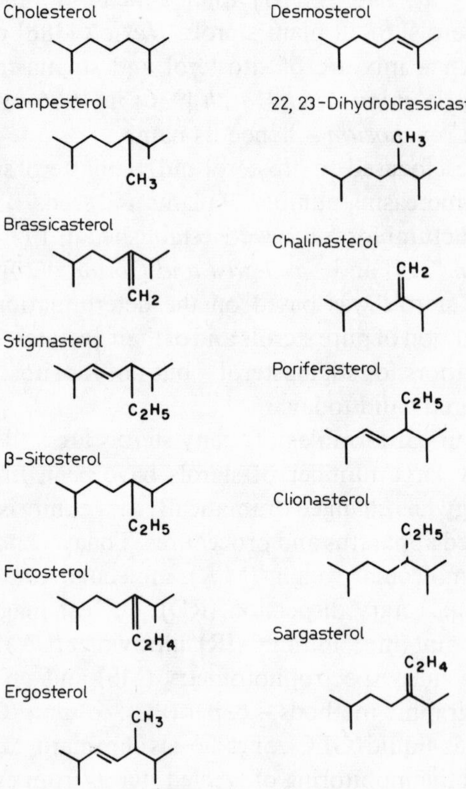

Cholesterol

Desmosterol

Campesterol

CH₃

22,23-Dihydrobrassicasterol

CH₃

Brassicasterol

CH₂

Chalinasterol

CH₂

Stigmasterol

C₂H₅

Poriferasterol

C₂H₅

β-Sitosterol

C₂H₅

Clionasterol

C₂H₅

Fucosterol

C₂H₄

Sargasterol

C₂H₄

Ergosterol

CH₃

Fig. 3. Common sterol side chains.

of these are given in figure 3. The empirical formulas of frequently encountered sterols and their epimers are presented (table I).

While it is expedient to use single-word designations for sterols the empirical and structural formulas must be kept in mind to appreciate the occurrence and location of double bonds and of alkyl groups, CH_3 and C_2H_5, since these characteristics play an important role in the penetration of cell membranes by sterols and thus in the absorption of sterols.

Table I. Empirical formulas of some important sterols

Cholesterol	$C_{27}H_{46}O$	cholest-5-en-3β-ol
		Δ^5-cholesten-3β-ol
Desmosterol	$C_{27}H_{44}O$	24-dehydrocholesterol
		24-dehydrocholesta-5,24-diene
Campesterol	$C_{28}H_{48}O$	24-α-methylcholesterol
		24-α-methylcholest-5-en-3β-ol
22,23-Dehydro-brassicasterol		24-β-methylcholesterol
Brassicasterol	$C_{28}H_{46}O$	24-α-methyl-3β-dehydrocholesterol
		24-α-methyl-Δ^{5,22}-cholestadien-3β-ol
		24-b-methyl-cholesta-5,22-dien-3β-ol
Chalinasterol		24-β-methyl-3β-dehydrocholesterol
Stigmasterol	$C_{29}H_{48}O$	24-α-ethyl-22-dehydrocholesterol
		24-α-ethyl-dehydrocholest-5-en-3β-ol
		24-b-ethyl-cholesta-5,22-dien-3β-ol
Poriferasterol		24-β-ethyl-22-dehydrocholesterol
β-Sitosterol	$C_{29}H_{50}O$	24-α-ethylcholesterol
		24-α-ethyl-cholest-5-en-3β-ol
		24-b-ethyl-cholesta-5-en-3β-ol
Clionasterol		24-β-ethylcholesterol
Fucosterol	$C_{29}H_{48}O$	24-ethylidenecholesterol
		20-β-methyl-24-methylene-cholesterol
		ergosta-$^{6,7,24(28)}$-trienol
Sargasterol		20-α-methyl-24-methylene-cholesterol
Ergosterol	$C_{28}H_{44}O$	24-β-methyl-cholesta-5,7,22-trien-3β-ol
		3β-hydroxy-24α-methyl-Δ^{5,7,22}-cholestatrien

3. Mislabeled Sterols

So far, three sterols have been shown to be mislabeled. Two are 'zoosterols' – bombicysterol from the silk worm, *Bombyx mori,* and muscasterol from the housefly, *Musca domestica.* A third is a 'phytosterol', γ-sitosterol, which has been found in many plants and seeds, in digitalis, corn oil, soybean oil, and in the secretions of several toads. Quite possibly, newer methodology will prove that other sterols are actually mixtures of two or more sterols.

Already in 1908–1910, *Menezzi and Morechi* [576] found that the sterol extracted from *Bombyx mori* contained 13–44% cholesterol. Then, in 1934, *Bergmann* [75] proved that the oil from *Bombyx* contained 33% sterols, of which 85% was cholesterol and 13% was sitosterol. In 1962/63, *Thompson et al.* [865, 866] disclosed that the γ-sitosterol from *Bufo vulgaris formosus* and also the γ-sitosterol from soybean and corn oil – previously considered to be a C_{24} isomer of β-sitosterol – are 50/50 to 75/25 mixtures of campesterol and β-sitosterol. *Morimoto* [596] and later *Sakurai et al.* [714] found that γ-sitosterol from toad cake (secretions) 'Ch'an Su' was a mixture of 93% cholesterol, 3% campesterol and 4% β-sitosterol. *Nishioka et al.* [617] reported that the γ-sitosterol from soybeans is a mixture of campesterol and β-sitosterol. *Linde et al.* [545], using GC and MS, found that the γ-sitosterol of *Digitalis canariensis isabellina* (Webb) is a mixture of three parts β-sitosterol, two parts campesterol and 0.7 parts cholesterol.

Of the mislabeled sterols, γ-sitosterol is of greatest interest to us since *Wilkinson* [927], the most outspoken and influential critic of the therapeutic use of sitosterol as cholesterol depressant, used γ-sitosterol in his clinical trials, whereas others used β-sitosterol.

B. Phytosterols in the Animal World

Phytosterols, singly or in sets of two or three, have been detected throughout the whole animal kingdom. For our purposes, invertebrates and vertebrates will be treated separately, and protozoa, helminths, insects, and marine vertebrates will be discussed in sections before we come to the higher vertebrates.

1. Invertebrates

a) Protozoa

Practically all members of the animal kingdom, including protozoa, require cholesterol or a closely related sterol to maintain life. However, more than other species, they seem to have selective needs for sterols of a specific structure.

Cailleau [158], in 1936, found that the growth of *Trichomonas columbay* can be supported by cholesterol, β-sitosterol, ergosterol, and other sterols. He concluded that the C_5/C_6 double bond was not indispensable (since dihydrocholesterol supported growth), that esterification of cholesterol did not suppress activity (since cholesteryl acetate supported growth), that prolonged hydrogenation did not matter (since cholestane was active), and that prolonged change in the double bond at C_1 did not matter (epicholesterol was active). However, side-chain break and replacement of one oxygen led to inactivity, as did degradation of the side chain.

Table II. Varied effect of dietary sterols on the growth of selected protoza. After *Hutner and Holz* [411]

Protoza	β-Sitosterol	Stigmasterol	Cholesterol
Labyrinthula vitellina var. pacifica	++	+	++
Paramecium aurelia var.4,stock 51.7(s)	++	+++	–
Trichomonas gallinae SLT	++	+	+++
Tetrahymena corlissi TR-X	+++	+++	+++
Tetrahymena paravorax RP	+++	+++	+++

+++ = Excellent; ++ = good; + = fair; – no growth.

The importance of the chemical structure for *Paramecium aurelia* var. 4, stock 51. 7s, was discussed by *Conner and Van Wagtendonk* [186]. A C_5/C_6 double bond seemed essential for growth. Esterification of 3-OH decreased activity, oxidation of the ring destroyed activity, presence of more than one double bond in the ring system inactivated the sterol, and so did dramatic change in the side chain. They found a number of sterols, including β-sito-, stigma- and brassicasterol, supporting growth; others, including cholesterol, did not. According to *Van Wagtendonk and Conner* [895] and *Van Wagtendonk et al.* [896], *Paramecium aurelia* required lemon juice for growth, and later they identified β-sitosterol as the growth factor in the juice. Since it promoted growth through successive transfers in concentrations of less than 50 n/ml, they suggested that β-sitosterol functioned as a coenzyme.

In 1962, *Hutner and Holz* [411] published a review (with 145 references) on the lipid requirements of microorganisms. Part of one of their charts is reproduced (table II).

Cadillos et al. [157] examined nine strains of *Entamoeba histolytica* and found that cholesterol inhibited one strain completely, five, partly, and three, not at all. Dihydrocholesterol, a by-product of intestinal bacterial action upon cholesterol, inhibited three strains completely and five, partially.

b) Insects
The sterol metabolism of insects, their ability to convert phytosterols to cholesterol, and the way diazasterols and other compounds interfere with such conversion are of particular interest – in spite of the obvious difficulties of extrapolating from insects to humans.

Table III. The effect of dietary sterols (50 mg) on the larvae of *Dermestis vulgaris.* After *Fraenkel et al.* [282]

	Cholesterol	β-Sitosterol	Ergosterol
Number of larvae	10	10	10
Number of pupae	5	1	0
Days of pupation	39.8	47	0
Days at death	0	18.2	10.4
Number of dead pupae	0	9	10

Selection of data is facilitated by two comprehensive reviews on the metabolism of steroids in insects, one by *Clayton* [175] (1964, with 156 references) and one by *Thompson et al.* [864] (1973, with 123 references). *Svoboda, Robbins* and *Thompson,* alone, together, or with others have, from 1963 to the present time, contributed well over 24 pertinent essays.

Phytosterol-Cholesterol Conversion. Cholesterol fulfills the dietary requirements of nearly all insects. According to *Clayton et al.* [176], the cabbage butterfly, *Pieris brassica,* can utilize only cholesterol. Another insect, the roach *Eurycotys floridana* Walker, absorbs cholesterol without esterification if the supply is high (500 μg), but esterification aides in the absorption when the supply is low (0.05 μg). Another insect not able to convert plant sterols to cholesterol is *Dermestes vulgaris,* bred by *Fränkel et al.* [282] on four different diets till pupation. The results indicated that cholesterol was irreplaceable (table III).

Curiously, a closely related insect, *Dermestes maculatus* de Gier *(vulpinus),* was not able to transform β-sitosterol to cholesterol but could convert it to desmosterol, which then became a 'sparing sterol', according to *Robbins et al.* [697].

Agarwal and Casida [5] found that the American cockroach, *Periplaneta americana,* converted cholesterol to other sterols. Its growth was inhibited and pupation delayed by β-sito-, stigma- and ergosterol. They interpreted the depression of the cholesterol content of larvae as the result of interference with cholesterol resorption by the phytosterols. In the absence of dietary cholesterol, partial growth and development were supported by β-sitosterol but not by stigmasterol.

Such selectivity is seen throughout the insect world, in phytophagous and in zoophagous insects. The German cockroach, *Blatella germanica,* is, according to *Clark and Bloch* [171] and others, able to modify ergos-

Table IV. Distribution of sterols (%) in the leaves of two varieties of pine and in the larvae of the sawfly. After *Schaeffer et al.* [722]

Sterol	Leaves *Pina virginiana*	Leaves *Pina rigida*	Larvae Neodiprion
β-Sitosterol	93	94	73
7-Dehydrocholesterol	0	0	6
Campesterol	7	6	4
Cholesterol	1	0	17

terol-$_{28}$C through oxidation, dehydration between C_{24} and C_{26}, and peroxidation. This roach cannot dealkylate β-sitosterol but is able to remove the C_{24} ethyl group, according to *Robbins* [695] and *Robbins et al.* [696]. *Wientjens et al.* [920] starved *Blatella* nymphae, then fed them β-sitosterol-3-^{14}C, plus 16 different antimetabolites in various amounts. Lastly, they fed a sterol-free diet. Extracts of the nymphae yielded varying amounts of cholesterol. On sitosterol feeding alone, 6% of sitosterol and 94% of cholesterol were recovered from the nymphae.

Bergmann and Levinson [74] tested 20 species of *Musca domestica* and found that the housefly is able to convert, at least partially, sitosterol to 7-dehydrocholesterol. *Kaplanis et al.* [443, 444] found that the housefly was unable to dealkylate β-sitosterol when reared on a semi-defined medium with 0.15% ^{3}H-β-sitosterol, but was able to utilize 7-dehydro-β-sitosterol without converting it to cholesterol. The ability of β-sitosterol was one fourth that of either cholesterol or campesterol. Only 1.4% of larvae reared on β-sitosterol emerged as adults and these failed to produce viable eggs. *Silverman and Levinson* [770] found that the only lipid in wheat bran fraction required by larvae of *Musca vicina* was β-sitosterol. *Brust and Fraenkel* [153] could maintain larvae of the blowfly, *Phormia regina,* with either cholesterol, ergosterol or sitosterol but not with zymosterol. *Schaeffer et al.* [722] provided data on the larvae of the Virginia sawfly, *Neodiprion pratti* Dyer. Its larvae obtained their sterols from the host plants, *Pinus virginiana* Mill and *Pinus rigida* Mill, by dealkylation of phytosterols to cholesterol (table IV).

Chalinasterol was extracted from the queen and the workers of the honey bee, *Apis mellifica* L., by *Barbier et al.* [50]. The bees accumulate sterols from pollen of the apple tree, *Pyrus mallus,* and the rock rose, *Cistus ladamiferus. Allais and Barbier* [20] and *Allais et al.* [21] fed ^{3}H-β-sitosterol to bees and found that cholesterol increased in the bees by the same amount by which sitosterol had decreased at the end of the study.

Experiments by *Ikekawa et al.* [413] are cited as an example of ester-ification of sterols during evolution. Larvae and pupae of the silk worm, *Bombyx mori* were injected with ^{3}H-β-sitosterol. Within 1 week, radioac-tivity had increased fourfold in the ether extract, which contained sterols, and had decreased by 20–25% in the benzene fraction, which contained steryl esters.

Cholesterol is not the only zoosterol found in phytophagous insects. *Svoboda et al.* [829] fed the Mexican bean beetle, *Epolachna varivestis* (Mulsant), on leaves of the Clark strain of the soy bean plant, *Glycina max* (L) Merr. This plant contained less than 1% cholesterol and nearly 98% of the three major phytosterols. In the eggs, pupae and adults, cholesterol was 2.9, 4.4 and 4.5%, respectively, and lathosterol was 1, 16, and 11%. This was caused by hydrogenation of the plant sterols.

Offhaus [631] introduced the rice flour beetle, *Tribolium confusum*, as a test subject. Applying different diets, he concluded that this insect has six dietary requirements for pupal development, two of which were choles-terol and ergosterol. *Fröbrich* [287] found that, for *Tribolium*, ergosterol could replace cholesterol in amounts as small as 1.5 µg/100 g food. *Svo-boda et al.* [828] reared newly hatched *Tribolium* larvae on a semidefined medium. Conversion rates to cholesterol and to 7-dehydrocholesterol were, respectively, 42 and 50% from ^{3}H-β-sitosterol, 47 and 53% from ^{14}C-desmosterol, and 41 and 50% from ^{3}H-stigmasterol.

Heed and Kircher [360] studied *Drosophila pachea* which breeds exclusively in the stem of the Seneca cactus, *Locopherus schotti*. In this plant there are two main sterols, Schottenol (Δ^7-stigmasten-3β-ol) and lophenol (4α-methyl-Δ^7-cholesten-3β-ol). Schottenol, whether as natural extract or synthesized, can replace the cactus proper and can sustain the fly. Lophenol cannot do this. A large number of sterols were studied. One of these, $\Delta^{5,7}$-stigmastadien-3β-ol, produced infertile females.

The tobacco hornworm, *Manduca sexta* (Johannson), was found to be well suited for steroid metabolism studies. *Hutchins and Kaplanis* [410] reared larvae for 24 h on 0.4% (dry weight) ^{14}C-4-cholesterol and unla-beled cholesterol. By GLC, 77% of the total sulfate fraction from meco-nium was shown to be cholesterol and 23% consisted of campesterol plus β-sitosterol. *Svoboda et al.* [824, 830] found that this oligophagous insect converts ^{3}H-β-sitosterol to cholesterol via $\Delta^{5,22}$-cholestadien and desmos-terol. Similar results with 22-dehydrocholesterol as diet suggested that Δ^{22}

is reduced before Δ^{24}, but that the presence of the Δ^{24}-bond is essential for the reduction of the Δ^{22}-bond. *Svoboda and Robbins* [825–827] determined the transition pathways for the three principal plant sterols: (a) β-sitosterol – fucosterol – desmosterol – 5,7,22-cholestatrien-3β-ol – 7-dehydrocholesterol – cholesterol; (b) stigmasterol – 5,22,24-cholestatrien-3β-ol – desmosterol – 7-dehydrocholesterol – cholesterol; (c) campesterol – methylenecholesterol – cholesterol. Conversion of phytosterols to cholesterol is by dealkylation of α- and β-methyl or ethyl, of methylene or ethylidene at C_{24}, or by dehydrogenation of double bonds at C_{22}.

The insects discussed above were selected to illustrate the various metabolic pathways which, to some extent, can be encountered in higher species.

Thompson et al. [864] suggested seven possible pathways in the utilization of sterols by insects: (1) esterification and hydrolysis; (2) reduction of double bonds; (3) oxidation, dehydration and peroxidation; (4) hydroxylation; (5) dealkylation; (6) side-chain cleavage, and (7) conjugation with sulfates. All these processes seem plausible if one considers the structural relationships between sterols and their conversion products.

Inhibition of Cholesterol Synthesis. The ability of many insects to transform phytosterols to cholesterol has been demonstrated. Initially, the pathways differ for the three main phytosterols, but they converge at the point of desmosterol formation. It is at this point that the process of conversion to cholesterol can be terminated by antimetabolites.

Svoboda and Robbins [826] found that combined feeding of phytosterols plus 22,25-diazasterol to *Manduca sexta* resulted in a disruption of cholesterol biosynthesis. For stigmasterol, the conversion stopped at, 5,7,22,24-cholestanetertraene-3β-ol. *Svoboda et al.* [830–833] singled out two substances, namely triparanol (MER 29) and the 3β-hydroxy-23,24-dinorchol-5-en-22-oic acid ('C_{23} acid', for short) as having the same effect as diazacholesterols. C_{23} acid fed to *Manduca* together with ^3H-β-sitosterol yielded, on analysis of prepupae, results which reflected the quantitative relationship between the dose of the antimetabolite (mg/100 g diet) and percent sterol distribution (table V).

It was concluded that C_{23} acid inhibits the reductase(s) which converts desmosterol to cholesterol. *In vitro,* inhibition was nearly complete; *in vivo,* desmosterol accumulated in the insect, apparently without ill effects. Since desmosterol decreased after pupation, the enzyme system seemed active beyond the larval state. When triparanol or diazacholesterol were

Table V. Relation between dose of 'C₂₃ acid' (mg/dl) and sterol distribution (%) in prepupae of *Manduca sexta*. After *Svoboda et al.* [833]

C₂₃ acid added	β-Sitosterol	Desmosterol	Cholesterol
26	27.4	54.3	18.3
13	26.1	49.7	24.2
10	20.3	29.2	50.5
5	23.3	26.4	50.3
1	15.7	13.4	70.9
0.5	23.3	7.4	69.3
Control	15.5	1.5	83.0

fed together with β-sitosterol, the latter was found unchanged in the tissues, blocking cholesterol available for ecdysone formation.

The significance of the above data lies in the fact that these substances have been recommended and used as hypocholesterolemic drugs. MER 29 was discontinued because of its many dramatic clinical side effects, rather than for its devastating interaction with lysosomes. The screening of drugs based on interference with cholesterol synthesis has been virtually abandoned.

Biologic Pest Control. Readers of this book will be surprised by the caption of this section which has no direct bearings on man's hypercholesterolemia and atherosclerosis. Yet, the topic should be of importance not only to ecologists, agronomists and economists but to all those interested in the environment.

Salama and El-Sharaby [715] found that addition of β-sitosterol to the medium renders larvae of the cottonleaf worm, *Spodoptera littoralis* Boisduval, sterile. *Heed and Kircher* [360] reported that $\Delta^{5,7}$-stigmastadien-3β-ol application produced infertile females of *Drosophila pachea. Guerra* [350], using a large number of chemosterilants to decrease female insect fecundity and thereby decreasing the insect population, found many chemicals which retarded larval growth of the tobacco bollworm, *Heliothis virescens* (F). His list included β-sitosterol. *Robbins* [695] noted that the number of larvae of *Musca domestica* reared on β-sitosterol is small and that adults are unable to produce viable eggs.

Because of their inhibitory ability, some phytosterols can be used in biologic pest control in the form of sprays of peach trees (against tree borers), of cotton (against boll weevil), and in chicken houses (against fly

Table VI. Range of concentration (in ppm) of five 25-azasteroids in the larval diet or medium required to kill or inhibit development in 75% of the test insects. After *Svoboda et al.* [832]

Larva	Sterol[1]				
	I	II	III	IV	V
Tobacco hornworm	2.0–4.0	0.75–1.0	0.25–0.5	0.5–0.75	0.1–0.25
Fall armyworm	4.0–8.0	4.0–8.0	0.75–1.0	1.0–2.0	1.0–2.0
German cockroach	>1,000	>1,000	>1,000	250–500	>1,000
Confused flour beetle	>1,000	100–250	50–100	250–500	>1,000
Yellow fever mosquito	2.5–5.0	0.5–1.9	0.5–1.0	0.5–1.0	0.5–1.0

[1] I = 3β-hydroxychol-5-en-24-dimethylamine;   II = 3β-methoxychol-5-en-24-dimethylamine;   III = 3β-cyclopentoxychol-5-en-24-dimethylamine;   IV = 5α-cholan-24-dimethylamine; V = 5β-cholan-24-dimethylamine.

larvae). Such sprays have the advantage of lasting more than a decade in one location.

Another approach to biologic insect control is the use of antimetabolites. *Svoboda et al.* [832] added five different 25-azasterols to the diet of five types of insect larvae and determined the concentration in the diet or in the medium which sufficed to kill the larvae or would inhibit the development of 75% or more of the insects (table VI).

Compound I had the highest potency. *Blatulla germanica* and *Tribolium confusum* had the greatest resistance to azasterols. This experiment indicated the inhibitory effect of azasterols on insect molting hormones and on metamorphosis. In contrast to the five species cited above as being sensitive, *Aedes* and *Musca* resisted these antimetabolites, probably because these species have an excessive supply of dietary cholesterol. This alerts us to the limits to which extreme hypercholesterolemia can be reduced in man.

c) Helminths

A brief account of nematodes seems in order here because of the ability of some of the helminths to transform sitosterol to cholesterol. Cholesterol is required for the growth of helminths.

Cole and Krusberg [183] used mp, IR spectra, TLC, and qualitative and quantitative GLC methods to study the sterol composition of two nematodes. *Ditylenchus triformis* and *Ditylenchus dispaci* contained cholesterol and lathosterol in 40:60 and 50:50 ratios, respectively. The first of these helminths had also a small amount of Δ^7-stigmastene-3β-ol and a fair amount of α-spinasterol, although its fungal host, *Pyrenochaeta terrestris,* has ergosterol as its major sterol.

Next, *Cole and Dutky* [182] worked with *Turbatrix aceti* and with *Panagrellus redivivus.* The first of these nematodes required at least 0.1 μg of cholesterol for growth. Several highly purified sterols supported growth, among these, β-sitosterol, fucosterol, chalinasterol, campesterol and stigmasterol. Sterols which differed in A and B rings could not replace cholesterol.

Dutky [241] and *Dutky et al.* [242, 243] studied the *DDl 36* nematode, a parasite of the wax moth, *Galleria mellonella* L. The nematode contained only 6 mg cholesterol and 13 mg Δ^7-cholestanol per 96.6 g weight. However, the host moth contained much cholesterol and only traces of campe-and of-β-sitosterol. This nematode can (a) convert 4-^{14}C-cholesterol by saturating the $C_{5,6}$ position and by rapid desaturation of the $C_{7,8}$ position to Δ^7-cholestanol, and (b) dealkylate ^{3}H-β-sitosterol and convert it to cholesterol. *Hieb and Rothstein* [387] found that cholesterol, 7-dehydrocholesterol, β-sitosterol and stigmasterol were able to support, to varying degrees, the reproduction of a free-living hermaphrodite nematode, *Caenorhabditis briggsae,* in a liquid nutrient with *Escherichia coli.*

d) Marine Invertebrates
This subregnum is of some interest because at least certain of its members are able to convert phytosterols to cholesterol. *Tamura et al.* [853] studied the metabolism of the small oyster, *Cressostrea virginica.* They fed one group a sterol-free diet for 32 days. Yet, their relative cholesterol content increased due to biosynthesis. A second group was fed β-sitosterol. This was digested and it augmented the endogenous sterol pool; however, cholesterol did not increase.

Kanazawa et al. [441] suggested that the prawn *Penacus japonicus* may be able to convert ergo-, stigma- and β-sitosterol to cholesterol.

Teshima [855] and *Teshima and Kanazawa* [856] fed sea-yeast tablets to crayfish, *Astacus astacus* and *Astacus fluviatilis,* to crabs, *Cancer pagurus* and *Portubus trituberculatus,* to lobsters, *Homarus gamarus* and *Panulirus japonicus,* and to prawns, *Penacus japonicus* and *Artemia salina. Penacus* prawns were injected with β-sitosterol and *Artemia* prawns with ^{14}C-yeast *Cryptococcus albus.* Both species were then treated with

^{14}C-cholesterol. In both prawns, radioactive cholesterol was obtained in pure form – by chromatography and/or by recrystallization to constant specific activity. C_{28} and C_{29} sterols were converted to C_{27} sterols by *Artemia salina* prawns.

2. Vertebrates

The presence of phytosterols in the animal kingdom has not been reported at great length, except in the insects. This lack of pertinent information may be ascribed to the lack of utilization of sophisticated methods which are available for the assay of sterols in animals and in humans. It could be argued that GLC, TLC and mass spectrophotometry (MS) are not needed in the clinical laboratory. However, it can also be argued that the methods commonly used for measuring cholesterol are inadequate, being neither sensitive nor specific, and that they have led to manifold erroneous reports and conclusions.

a) Non-Human Vertebrates

In 1969, *Minato and Otomo* [594] analyzed the body oil of *Chelonia mydas mydas* Linné, the green turtle of New Guinea. It contained 96% cholesterol, 3% β-sitosterol, 0.2% campesterol, and 0.18% stigmasterol. Studies of the green turtle of Costa Rica yielded similar results. Two years earlier, *Miettinen* [585] subjected extracts of the adrenal glands of rats to chromatography on silica gel and silver nitrate-impregnated silica gel and to GLC. Three unsaturated $\Delta^{5,7}$-sterols, campe-, stigma- and β-sitostanol, constituted about 20% of total sterols, and cholestanol accounted for 80%. Of the assorted α-sterols, campe-, stigma- and β-sitosterol constituted 5%, the same as in the diet, whereas cholesterol amounted to 94% and dihydrocholesterol to 1%. The adrenals of rats subjected to stress at low temperature had markedly lower cholesterol, a mild decrease in dihydrocholesterol, and no change in the three phytosterols.

Boorman and Fisher [134] found β-sitosterol and campesterol in the blood plasma, intestinal wall and liver of adult hens fed maize sterols. More campesterol than β-sitosterol was present.

Nes et al. [609] compared normal rats with tumor-bearing rats. Campesterol and β-sitosterol were isolated as 1% of total liver sterols. There was no difference between normal livers and livers with Morris rat hepatoma. When rats with hepatoma were injected with 22,23-^{3}H-24-ethylcholesterol, 15% of the label was distributed between tumor tissue, intact liver tissue, and the rest of the body. The 24-ethyl label was not found in any of

Table VII. Sterols present in human cancer tissue. After *Day et al.* [217]

Case No.	Age	Primary neoplasm	Site	Metas-tases	Sterol[1] CH	D	S	CA	β
1	56	papillary adenocar.	kidney	+	++	±	–	–	–
2	58	ductal cell car.	breast	+	++	±	–	–	–
3	42	ductal cell car.	breast	+	++	±	–	–	–
4	58	adenocar.	breast	+	++	±	–	–	–
5	24	granulosa-theca car.	ovary	–	++	±	–	–	–
6	38	adenocar.	breast	–	++	±	–	–	–
7	36	adenocar.	breast	+	++	±	±	–	–
8	49	follicular car.	thyroid	+	++	–	–	–	–
9	65	mullerian car.	ovary	–	++	±	±	±	–
10	32	medullar car.	breast	–	++	–	–	–	–
11	49	medullar car.	breast	+	++	–	–	–	–
12	45	car.	breast	+	++	±	–	–	–

car. = Carcinoma;   adenocar. = adenocarcinoma;   ++ = Large   amount;   ± = very   small amount; – = none.
[1] CH = Cholesterol; D = desmosterol; S = stimgasterol; CA = campesterol; β = β-sitosterol.

the sterols. The presence of β-sitosterol in the liver was thought to be of dietary origin since feeding experiments led to comparable results.

b) Man

With Neoplasm. In recent years there have been several reports on phytosterols in human neoplasms and in the blood serum of cancer patients. *Day et al.* [217] analyzed the sterols in cancer tissue of 12 patients. All were blacks, all but one were females, and all had been operated on 1 month to 2 years previously (average, 6 months). There was no correlation between the patient's age, the site of neoplasm, its spread, duration, treatment, serum Ca, K, and alkaline phosphatase, and the presence, amounts and types of five sterols. Cholesterol was found in all tissues, desmosterol in 10 out of 12. Phytosterols were found twice in primary cancers and twice – in two other patients (1 and 8) – in the metastasis (table VII).

About two thirds of cancer patients with hypercalcemia have breast cancer with osteolytic metastasis. One third have bronchogenic carcinoma, parathyroid adenoma, or other neoplasms – without osteolytic metastasis. Because of this, *Gordan et al.* [319] implanted micropellets of vitamin D or of cancerous breast tissue against the skull of rats. This resulted in osteolysis. They searched for parathyroid hormone in extracts of breast cancers of 12 women. They found none in the aqueous fraction. The lipid fraction, however, was highly osteolytic. On TLC, there were four bands: the second band contained most free sterols, the fourth, most steryl esters. The first and third fractions had no osteolytic activity. The fourth fraction contained, on GLC, the three phytosterols. In plasma extracts of these patients the three plant sterols were also present, especially stigmasteryl ester.

The obvious question was asked, whether cancerous tissue is able to synthesize phytosterols or whether these are of dietary origin. On injection of 250 µg on 2 consecutive days, the average urinary ^{45}Ca, in dpm/day, was 1,320 on ergosterol, 630 on stigmasterol, and 2,640 for the osteolytic stigmasteryl acetate. *Gellhorn,* discussing the findings of *Gordan et al.* [319], asked why some patients with hypercalcemia have osteolytic activity and some do not. Synthesis of phytosterols by cancerous tissue seems unlikely since these sterols have been detected in normal tissue. In the plasma of 16 healthy women the three main phytosterols were present as often as in the blood plasma of 19 women who had disseminated breast cancer. Esterified phytosterols were infrequent in healthy subjects and, when present, were represented by stigmasteryl esters only.

To pursue the question further, *Huddad et al.* [408] studied, using GLC, 11 healthy women, 7 with lactating breasts, and 14 with mammary carcinoma. No significant differences in plasma phytosterols were found in the three groups. No stigmasteryl acetate was found. Cancer patients with eucalcemia had sterol values comparable to normal subjects. In 1 patient with hypercalcemia the plasma phytosterol levels decreased during a hypercalcemic period. In another patient the level changed after removal of the carcinoma. The phytosterol levels were not typical of lactation or of cancer, nor could they be held responsible for the hypercalcemia of cancer patients. Of the 14 women with breast cancer, 10 had metastases, 4 had none. Yet all had similar plasma sterol levels. Data from the above study was as follows: the three phytosterols were found in amounts of 7,320 n/g cancer tissue, 7,709 n/g normal breast tissue, 6,533 n/g fat, and 49,934 n/g bile. Mean plasma levels always correlated with food consumption.

Osteolytic phytosterol acetates were the subject of a GLC study be *Whitney et al.* [919]. Serum campe-, β-sito-, and stigmasteryl acetates of 14 patients with breast cancer were 2.7, 5.9, and 5.6 n/ml, respectively. Their sum was 13.2 n/ml. In 4 women with benign breast lesions, the corresponding amounts were 2,3, and 1 n/ml. The total was 5 n/ml. In 9 healthy women, the total was 35.6 n/ml, and in 9 healthy men, the total averaged 74.1 n/ml. The phytosterol levels of healthy subjects, small as these levels were, exceeded those of patients with benign breast lesions.

Grossi-Paoletti and Paoletti [343] reported on the occurrence of desmosterol in human glioblastoma and oligodendroglioma, but not in astrocytomas or non-glial tumors.

Without Neoplasm. As is evident from the preceding paragraphs, some studies investigate patients with, and patients without, neoplasm. References to plasma phytosterol levels in healthy persons are mentioned in this section and will be considered again in the frame of discussion on sterol absorption. A certain amount of repetition is unavoidable.

In 1964, *Böhle et al.* [133] studied the resorption of phytosterols consumed by 18 healthy persons and by 13 patients with atherosclerosis and hypercholesterolemia-hyperlipidemia. Using GLC, they found small amounts – maximum, 4.74 mg/dl – of β-sitosterol and stigmasterol in the blood of all 31 subjects. The amounts were 6.66 times greater in healthy persons than in those with atherosclerosis. *Gray et al.* [340] found, by TLC, GLC, and GC-MS, 10–40 μg/dl of free or esterified β-sitosterol and 10–70 μg/dl of free coprostanol (cholesterol-α-oxide) in extracts of human serum.

Lees and Lees [529] detected, on ingestion of 3 g sitosterol/day, 0.47–2.07 mg/dl, and on ingestion of 6 g/day, 0.63–2.45 mg/dl in plasma. In 18 subjects, following intake of 18 g 'Cytellin' (65% β-sitosterol), they found 0.37–0.75 mg/dl β-sitosterol (average, 0.48 mg/dl) and 4–21 mg/dl campesterol (average, 16.8 mg/dl). In infants and children, *Mellies et al.* [572] found considerable differences in plasma phytosterols, which reflected the diets. Those on a low-cholesterol/high-phytosterol diet had substantially higher levels (table VIII).

Accumulation of plant sterols in cancerous tissues has not been established. Their accumulation in healthy tissues is negligible – if it occurs at all. Phytosterols in small amounts are found in the blood plasma of many, if not of all, persons who consume plant sterols.

With Hypersitosterolemic Xanthomatosis. The patients to be discussed here differ from all the others. In sharp contrast to the minute amounts of phytosterols in the plasma and tissues of healthy persons and

Table VIII. Phytosterols (mg/dl) in sera of infants and children. After *Mellies et al.* [572]

Number	Subjects	Diet	Sterols[1]			
			CH	CA	β	total
15	infants	breast or whole milk	0.70	0.69	0.83	2.22
9	infants	formula	2.45	0.72	3.53	6.70
37	children	normal diet	0.80	0.42	0.31	1.53
14	children	normal	0.69	0.20	0.69	1.58
20	children	low cholesterol/high phytosterol	2.55	1.61	1.83	5.99

[1] CH = Cholesterol; CA = campesterol; β = β-sitosterol. Subjects with type II familial hypercholesteremia.

those with carcinoma or with other diseases, these patients have a high phytosterol level in their blood, especially, large amounts of β-sitosterol. As in many patients with essential hypercholesterolemia those with hyper-β-sitosterolemia suffer from xanthomatosis.

1973–1975, *Bhattacharyya and Connor* [123–125] published their observations on 2 sisters, L. H., aged 23, and R. H., aged 21 years, who had familial xanthomatosis. L. H., whose plasma cholesterol level was 203 mg/dl, also exhibited 37 mg/dl of phytosterols (27 mg of β-sitosterol, 10 mg of campesterol) while her sister, R. H., whose plasma cholesterol level was almost identical (206 mg/dl), showed 26 mg/dl of phytosterols (17 mg β-sitosterol, 8 mg campesterol). Both had 0.5 mg/dl of plasma stigmasterol. The authors made two reports on the sterol distribution in the xanthomatous lesions. Their detailed metabolic experiments showed that hypersitosterolemia was due to enhanced absorption of this sterol. Analysis of the parents' blood suggested an inherited, recessive trait.

A third instance of β-sitosterolemia and xanthomatosis was reported by *Schulman et al.* [752]. In a 31-year-old woman, onset of the disease had occurred at age 1½, and she fully developed xanthomatosis at age 13. Her plasma cholesterol fell from 310 to 250 mg/dl and her low-density LP (LDL) cholesterol from 225 to 160 mg/dl on a polyunsaturated fatty acid (PUFA) diet; the two levels fell to 190 and to 100 mg/dl, respectively, on 16 g/day cholestyramine for 8 months. However, xanthomata increased. The patient's plasma cholesterol was 242 mg/dl, phytosterol 18 mg/dl (12 mg β-sitosterol, 6 mg campesterol, and traces of stigmasterol) and analysis of xanthomata yielded 8.3 mg/100 g cholesterol, 1.5 mg/100 g phytosterols (1.34 mg β-sitosterol, 0.8 mg campesterol, and traces of stigmasterol).

The plasma phytosterols of these patients have a dietary origin. An underlying hereditary enzymatic defect seems likely. If, to date, no other patients with phytosterolemia have been reported, it may be due to lack of GLC studies of patients with xanthomatosis.

c) Phytosterol-Cholesterol Conversion by Vertebrates

The observations that insects and prawns are able to transform phytosterols to cholesterol prompted a search for analogous events in higher species.

Some investigators believe that the ingestion of β-sitosterol influences cholesterol biosynthesis by man. This may occur in cases of excessive intake of β-sitosterol and it may represent a homeostatic regulatory process. To date, it has not been proven that phytosterols can be the source of endogenous cholesterol. At least two species, pigeons and rats, are, however, able to oxidize not only cholesterol but also the three main phytosterols to their respective stanols. They can also dealkylate phytosterols.

Subbiah et al. [808] found in the course of extensive GLC study of pigeon metabolism that α^5-stanols were converted to Δ^5-sterols by the birds. Few β-stanols and β-ketones were present in the feces. *Evrard et al.* [261] followed the process of oxidation of OH at C_3, producing keto-derivatives, and the saturation of double bonds at C_5, C_6, producing the homologues – the stanols, when rats were fed any of the three phytosterols. In rat feces, β-stanols and β-ketons were plentiful.

Rat and mouse liver mitochondria were used in studies by *Kritchevsky et al.* [500]. Ergosterol-29-^{14}C was readily oxidized to $^{14}CO_2$ *Subbiah and Kuksis* [812] used a lecithin emulsion which contained $1 \mu Ci$ of 4-^{14}C-β-sitosterol ($10 \mu Ci/mg$) for incubation of mitochondria, plus cofactors required for cholesterol oxidation. In 16 h, 7–10% of sterols were oxidized by 2 g liver tissue. Mono-, di- and trihydroxy bile acids were formed. Rat liver mitochondria converted 10% of β-sitosterol – to stigmasterol! The major oxidation products of β-sitosterol and cholesterol were analogous, but relative amounts differed.

Subbiah [801] injected rats intravenously with a lecithin emulsion of sitosterol-4-^{14}C. Its tissue distribution was the same as that of cholesterol. *Subbiah and Kuksis* [811] collected bile for 4 h, drained the blood, then removed the organs from rats. Of the ^{3}H label, 96% was in the liver, 1.8% in the bile. The rest was in erythrocytes, adrenals, kidneys, heart, and testes.

Swell and Treadwell [844] injected i.v. cholesterol-4-^{14}C or phytosterol-^{14}C into normal rats and into rats with bile fistula. 24 h later, about 20% of either label appeared in the liver. Serum, liver, small intestine and adrenal cortex contained some sitosterol, but much less than cholesterol. Less of the sitosterol was esterified. The bile contained more cholesterol than sitosterol. This was interpreted as selective secretion of sterols by the liver. It was concluded that there is no conversion of phytosterols to cholesterol or related sterols.

Kuksis and Huang [517] reported that the rat converts sitosterol to coprositostanol and that cis-stanols are also produced from cholesterol, β-sitosterol, and campesterol. When added to the diet in amounts of 50–100 mg, all the sterols are reduced at equal rates.

It is interesting that *Goad et al.* [313] and *Goad and Goodwin* [314] found a similarity between metabolism of mevalonic acid by rat liver homogenates and larch. *Svoboda et al.* [833] reported that 'C$_{23}$ acid' has a comparable effect on larvae of *Manduca sexta* and on rat liver tissue *in vitro.*

C. Zoosterols in Plants

The number of 'zoosterols' found in plants is still small. It is bound to grow. The first report on cholesterol in plants was made in 1957 by *Tsuda et al.* [879]. In red algae *(Rhodophyceae)* they identified chole-, sito-, sarga-, fuco- and chalinasterol. They also found cholesterol in five plants of the genus *Gelidium. In 1959, Boehme and Völker* [132] identified cholesterol, by UV absorption and PC, next to large amounts of β-sitosterol, in the peels of California oranges. *Reitz and Hamilton* [689] studied the blue-green algae, *Anacystis undulans* and *Fremyella diphsiphon.* The sterols in these algae were tightly bound but, after tedious separation, the β-sitosterol:cholesterol (S:C) ratio was identified as 50:50. Some stigmasterol was present and traces of other sterols. The principal sterols were apparently synthesized by the algae.

Gawienowski and Gibbs [294] recovered cholesterol from apple seeds, *Knights* [486] found small amounts of cholesterol in oats, and *Ardenne et al.* [31] obtained it from potatoes. Multiple references have been to the potato plant, *Solanum tuberosum,* as a source of cholesterol. Some of these are cited here. *Johnson et al.* [422] identified cholesteryl, β-sitosteryl and stigmasteryl acetates by GLC, TLC, mp, and by purification to constant radioactivity, in stems and leaves of potato plants. *Schreiber and Osske*

[748] extracted from the leaves, besides β-sitosterol, 4α-methyl-5α-stigma-7,24(28)3β-ol which they thought to be identical with α-sitosterol. All these sterols could find their way into the Colorado potato beetle, *Leptinotarsa decilineasta,* say, whose principal sterol is β-sitosterol, according to *Schreiber et al.* [749].

The finding of cholesterol in various plants, especially in potatoes, should be of interest to those who attempt to influence man's blood cholesterol by a vegetarian diet or by a diet enriched by potatoes, such as have been recommended lately.

1. Sterols in the Flora

a) Distribution
The task of writing this chapter is aided by several excellent reviews. In 1961, *Crombie* [206] listed 24 sterols and their distribution in plants. *Heftmann* [361] enumerated in the year 1963 plants with 51 different sterols and steroids: 20 sterols, 6 saponins, 6 alkaloids, 11 cardiac glycosides, 5 pregnane derivatives, and three hormones. *Shoppee* [766] listed in 1964 the sources of 26 sterols and pointed out the overlap between phytosterols, mycosterols, and marine sterols. *Bean* [56], in his review of 1973 (with 112 references), enumerated 44 sterols and their distribution in seven classes of plants. The greatest number, 23 of 44 sterols, was in angiosperms. Cholesterol was found in 20 types of angiosperm. As to frequency, β-sitosterol was listed 42 times, stigmasterol 31 times, and campesterol and cholesterol each 29 times. This places cholesterol among the 'phytosterols'.

The list of medicinal herbs and ornamental shrubs from which sterols have been isolated is growing steadily. Since modern instrumentation has become available, the isolation and identification of sterols in the flora has become a favorite topic for essays and for doctoral theses.

Sterols can be found in the most primitive and in the highest animals and plants. There is a blurring between zoo- and phytosterols. There is also an overlap between zoo- and phytoflagellates, and coexistence of phyto- and zooplankton. References to plankton and humus, and to bacteria and molds, will be separated from those to the higher flora.

b) Plankton and Humus
Plankton and humus, which, in some ways, is a counterpart to plankton, are examples of the overlap of flora and fauna. *Baron and Boutry* [51] analyzed Mediterranean plankton and identified cholesterol, β-sitosterol,

stigmasterol, lanosterol, and 7-dehydrocholesterol. *Boutry and Baron* [139], on replacing PC with spectral analysis, found 88 and 96% cholesterol, 5 and 2% campesterol, and 4.5 and 0.7–1.5% β-sitosterol, in the Monaco Sea and the Vezina Lake, respectively. Sunlight played a critical role in the distribution of sterols. *Matthews and Smith* [560] studied the ratio of cholesterol to β-sitosterol in plankton and marine waters in the Gulf of Mexico. The ratio was inversely proportional to the distance from the shore. *Henderson et al.* [372] examined the Mono Lake in California. Its sediment, with alternate bands of algal silt and diatomite, representing in a complete Pleiostocene lacustrine sequence 130,000-year-old deposits, yielded a large variety of zoo- and phytosterols.

Humus contains, according to *Cerbulis and Taylor* [161], 1.2–6.3% of lipoid from non-decomposed plant residues and from the bodies of living and dead microfauna. This explains the presence of cholesterol, β-sitosterol, γ-sitosterol, stigmasterol, campesterol and ergosterol in the unsaponifiable matter extracted from the earthworm, *Lumbricus terrestis.*

c) Bacteria and Fungi

The number of essays on sterol metabolism of microorganisms is very large. Only a few references are selected in order to illustrate the investigative approaches, to point out some deficiencies of studies on the sterol metabolism of bacteria, and to underscore the observations which have a bearing on the sterol metabolism of man.

In 1958 and 1962, *Bergmann* [77, 78] contributed two provocative reviews on the evolutionary aspects of sterols. He labeled cholesterol the 'fittest' sterol which survived in the phylogenetic evolution from bacteria and protozoa to higher plants and mammals. He referred to β-sitosterol as a cholesterol homologue in the chemical and also the biologic sense.

Smith [776, 777] studied pleuropneumonia-like organisms which require cholesterol or cholestan-3β-ol for growth. Other sterols, though incorporated into the organisms, could not support growth. Some important conclusions evolved from these studies: there is an absolute requirement for a planar molecule with an equatorial hydroxy group; the sterols must have a highly specific configuration, namely the radical in equatorial (not in axial) position to be able to resist hydrolysis; sterols have to penetrate the cell membrane before they can be utilized.

Schubert and Rose [751] made an extensive study of *Escherichia coli* and found, using GLC at two different evaporation temperatures, four and six components, respectively. The crystalline mixture contained 25%

cholesterol, 23% β-sitosterol, 21% campesterol, 18% stigmasterol, 8% dehydrocampesterol, and 4% of an unidentified sterol.

Rosenfeld and Hellman [704] incubated feces with cholesterol-1,2-^{3}H and sitosterol-4-^{14}C. They found that the structural differences between these two sterols were sufficiently remote from the A-ring to have any effect on reduction or on esterification. Bacterial destruction of the sterols was the dominant activity and there was little esterification. Equal reduction of Δ^5-double bonds to the saturation analogues, coprocholestanol and coprositostanol, resulted.

According to *Coleman and Baumann* [184], rat fecal bacteria grown on sterol-free media converted β-sitosterol first to coprositostanol, then to 7-dehydrocholesterol and coprost-7-enol. On addition of 50 mg sitosterol to the media, saturated sterols increased to 60–80% of the added amount; sitosterol reduction was 67–69%.

Preferential uptake of campesterol by chickens – when compared to pigeons, rodents, and humans – is ascribed by *Subbiah and Kottke* [807] to a difference in the bacterial flora, being able to transform Δ^5-sterols.

Subbiah et al. [815] studied 10 patients with hyperlipoproteinemia. Sterol absorption decreased in the order $C_{27} > C_{28} > C_{29}$. The ratio of fecal cholesterol to phytosterols varied with bacterial degradation of the sterols. It was not the same in all 10 patients. Supplementary studies of fecal homogenates revealed a lower conversion rate of [^{3}H]-β-sitosterol to 5β-stanols, although the recovery of the two sterols was comparable.

Reference to the role of intestinal bacteria in the degradation of neutral 3-β-OH-Δ^5-sterols to products which are not readily recognized as steroids in the analysis of human feces was made by *Grundy et al.* [346]. The loss of ^{14}C-cholesterol and β-sitosterol-^{3}H amounted to 28–50% of cholesterol and to 23–27% of β-sitosterol – with the lower figures following instillation of high doses, the higher figures following instillation of low doses. These observations point to the importance of quantitative relationships of dietary sterols and their influence on the relative absorption of sterols and the degree of their conversion to stanols.

Den Besten et al. [223] observed 6 men who had been placed on a formula diet. In 5, there was 100% recovery of dietary sitosterol in the feces. In a sixth man, only 25–58% of ingested sitosterol was recovered: when powdered fresh celery or cellulose was added to the formula, recovery of sitosterol increased to 80%. Lactose, added to formula, increased recovery to 58–80%. Addition of cellulose plus lactose resulted in 100% recovery of sitosterol.

The role of cellulose and/or fiber, with its ability to adsorb sterols and possible ability to interact chemically with sterols, has become a popular research topic. Vegetarians, lactovegetarians, and lacto-ovo-vegetarians form good study groups: the influence of these diets and also that of crude fiber on the intestinal microflora has not been sufficiently studied to date.

The differences attributed to intestinal bacteria must be borne in mind when one tries to interpret the differences in sterol metabolism and the incidence of atherosclerosis in various ethnic groups and at various locations. Again, the reports which implicate the intestinal flora lack data on the type of microorganisms. Epidemiologic data, clinical observations, biochemical and bacteriologic studies should be correlated because of the obvious preventive-therapeutic implications.

Sterols are needed for growth and reproduction in fungi, although there are exceptions to this rule. *Trichophytum rubrum, Aspergillus niger, Collotrichum lagnarium,* and *Pericularia orizea* can grow in the absence of sterols. However, the presence of sterols does stimulate growth. The mycelium of *Phythium* (PRL 2142) was best stimulated by an active fraction of sunflower seed oil – β-sitosterol. The specific requirements for growth were the length of the C_{17} side chain, the stereochemical configuration, and the position of the oxygenated fraction. *Haskins et al.* [358] reported that many species of *Phythium* and *Phytophthora* did not form sexual bodies in synthetic media but required addenda, such as β-sitosterol and cholesterol, each of which was effective in 0.2 μg/ml of potato-dextrose agar. *Knights* [486] confirmed that the fungus *Phytophthora cactorum* failed to produce mature oospores in a basal medium, but did so when grown on oat meal which contains β-sitosterol and other sterols.

Of the large number of mycosterols, or fungisterols, only ergosterol, because of its relation to vitamin D, has attracted wide attention.

2. Factors Affecting the Distribution of Sterols in Plants

There are considerable differences in the amounts of sterols in plants – from one class to another, between members of the same genus or family, and from crop to crop. The partition of sterols varies between various parts of a plant, often with photoperiods or with seasons. Some quantitative differences exist between cell membrane, mitochondria, and lysosomes.

Korytek and Metzler [490] analyzed beans. All beans are rich in linoleic acid, as reflected in a high iodine number of extracts. The glyceride fraction of extracts contains palmitic and linolenic acid, and some stearic and oleic acid. The differences in total lipid are not striking; their range is 0.9–1.95%. However, unsaponifiables are as low as 6% and as high as 30%. Thus, diet will be affected by the kind of beans eaten: kidney beans had 6%, pinto beans 7%, blackeye beans 10%, and small California beans had 19% phytosterols. Moreover, the 1958 crop of lima beans had 19% of unsaponifiables, the 1959 crop had 30%. Seasonal variations were not significant but annual variations were substantial.

Another example of differences – this time in the relative amounts of seven sterols – in material from closely related sources was reported by *Nagy and Nordby* [604]. They analyzed eight varieties of orange and tangor juices and found considerable variations (50–400%) in their sterol content.

Light affects the content and distribution of sterol in plants. Autooxidation of sterols in the plankton is influenced by the ozone, aeration and sunlight. This applies also to higher plants. *Bae and Mercer* [43] reported that the leaves of *Solanum audigena* contained less β-sitosterol and cyclosterol during long-day photoperiods. Cholesterol decreased during short-day periods. Generally, leaves contained less sitosterol than stems. In some instances this distribution may be reversed due to alkalinity of the soil, according to *Misra et al.* [595]. They added powdered extract of *Argemona mexicana* to the soil and found that the stem contained 0.03% and the leaves 0.06% of β-sitosterol.

Davis [215] reported that stigmasterol increased and β-sitosterol decreased with the shedding of leaves, and that campesterol and cholesterol increased slightly from the base to the top of the tobacco plant. During maturation, total sterols decreased while the percent of constituents remained stable. Here, we have a combination of the light factor with the growth factor, both altering the sterol distribution.

The sterol content of various parts of plant cells differs. *Kemp et al.* [454, 455] and *Kemp and Mercer* [456, 457] studied various parts of corn seedlings after germination. Using seedlings of different age, they found that sterols increased with aging in shoot and root but remained the same in other parts of the plant; Δ^5-sterols increased only in the scutellum, not elsewhere. In root tissue, stigmasterol increased, β-sitosterol decreased. The bulk of the sterols, however, was found in the mitochondria and microsomes. The nuclear fractions contained the most esterified sterols, and the microsomal fraction the least. Most of the esterified cholesterol was in nuclear and chloroplast fractions, the nuclear being the only one with an appreciable amount of free cholesterol.

Ingram et al. [414] stated that in the genus *Crucifera* β-sitosterol is synthesized in the growing plant, i.e. after germination. They found great differences in the amounts and in the distribution of sterols in seven related species.

All these variables are presented for consideration by those who attempt to influence cholesterol metabolism by dietary manipulation. One must take into account the vagrancy of nature.

3. Biosynthesis of Sterols in Plants

Several excellent reviews on the subject of sterol synthesis in plants are available. One, by *Bergmann* [76], has 493 references. Two others are by *Heftmann* [362, 364] with 190 and 214 references, respectively. Only selected essays, therefore, are cited here.

Comparison of the structural formulas of sterols elucidates the conversion routes and biosynthesis, which is similar to sterol synthesis in animals. The concept of sterol synthesis by plants is strengthened by the fact that cholesterol is present in many plants and, moreover, in all parts of plants, from seeds to pollen.

Benevista et al. [67] treated *in vitro* tissue cultures of tobacco leaves with ^{3}H-acetic anhydride and followed the progression to cycloartenol, methylene-24-cycloartenol, cycloeucalcenol, obtusiofoliol, 'lophosterols', and on to either β-sitosterol, campesterol, or stigmasterol. *Goad and Goodwin* [314] observed incorporation of [2-^{14}C]-mevalonate into squalene and into the three main phytosterols of pea and of maize leaves. The presence of cycloartenol (by GLC), an early conversion product of squalene, and of fucosterol, a late product, is noteworthy. Incorporation of the label into squalene was drastically reduced under anaerobic condition; labeled squalene accumulated. *Goad* [312] pointed to fucosterol as the last intermediate in the process of (CH$_3$) alkylation during biosynthesis of C$_{28}$ and C$_{29}$ sterols. This is the reverse of the dealkylation observed in insects, where fucosterol precedes desmosterol in the conversion of phytosterols to cholesterol.

Waters and Johnson [905] grew *Glycina soja* to a height of 3–4 inches in daylight or under UV for 24 h and incubated the plants with either *DL*-mevalonic acid-2-^{14}C or *DL*-[methyl-^{14}C]-methionine. They isolated radiochemically pure stigmasterol and β-sitosterol from the leaves of the plants. Their conclusion that no biosynthetic relationship exists between the two sterols – because none was detected via hydrogenase or dehydrogenase – may be incorrect.

Table IX. Distribution of ^{14}C from ^{14}C-mevalonic acid (mg/g) in various parts of *Salina sclerea*. After *Nicholas* [611]

Source	Sclareol	β-Sitosterol
Flower	39.19	3.09
Flower stems	26.18	1.21
Leaves	2.75	1.36
Main stem	4.18	0.58
Stipules	23.81	1.45

The findings were contradicted by *Bennett and Heftmann* [68] and by *Bennett et al.* [69] who reported conversion of β-sitosterol to stigmasterol by leaf cuttings of $3^{1}/_2$-month-old potted *Digitalis lanata.* In *Discorea spiculiflora* plants this conversion was clearly due to dehydrogenation of β-sitosterol.

Labeled mevalonic acid has been used repeatedly as starting material for the study of phytosterol biosynthesis. *Johnson et al.* [423] found maximum incorporation of mevalonic acid-2-^{14}C by the Kathadien variety of *Solanum tuberosum* into β-sito- and stigmasterol. *Goad et al.* [313] incubated leaves of larch, *Larix decidua,* with $2R[2-^{14}C_1(5r)-^3H]$-mevalonic acid. The resulting $^3H:^{14}C$ ratio was 6:6 for cycloartenol and 6:5 for β-sitosterol. Formation of C_5, C_6 double bonds of phytosterols also involved the elimination of the 6α-H atom from precursor sterols.

Salina sclerea of the family *Labiata* incorporated the label of 2-^{14}C-mevalonic acid, according to *Nicholas* [611]. The incorporation was preferential into sclareol, i.e. 12.5 times greater than into β-sitosterol (table IX).

Baisted et al. [44] studied the biosynthesis of sterols in germinating *Pisum sativum* seeds. During a 5-day period, 45% of the radioactivity of 2-^{14}C-mevalonic acid was incorporated into β-amyrin, but only 2.2% into β-sitosterol. The latter required a longer and more complicated process: (a) partial cyclization of squalene; (b) external alkylation; (c) reduction of the terminal double bonds; (d) removal of three methyl groups, and (e) exchange in the unsaturation of ring B.

Incorporation of precursors into the side chain of various plant sterols has been studied by *Mercer* [578], *Rees et al.* [683, 684], *Castle et al.* [160], and *Heftmann* [362, 364]. The possible modes of alkylation have been discussed by *Lederer* [528]: (a) C-methylation from methionine; (b) incorporation of propionic acid, or (c) leucine-isoleucine, or (d) of mevalonic acid.

4. Function of Sterols and Steroids in Plants

The functions of sterols and steroids in plants parallel the functions of these compounds in animals. *Cook* [194, 195] and *Heftmann* [366, 367] have reviewed the topic in detail. In summary, their roles are defined as follows: (1) in the relation of sterols to cell membrane structure and permeability, (2) their function as steroid hormones; (3) the steroid-hormonal activity of phytosterols; (4) their function as hormone precursors, and (5) as source of insect molting hormones. Possibly, the only different sterol function is the anti-inflammatory action of cholesterol in animals.

A statement made in 1930 by *Schönheimer er al.* [744] that 'plant sterols are just waste products' is of some historic interest.

a) Sterols in the Cell Membrane

The biochemical and biophysical architecture of plant cell membranes resembles that of animal cell membranes. The controversies concerning the differences between monolayers, condensed monolayers, mixed phospholipid monolayers and lipid bilayers apply to both plant and animal cells. So do the different resorbabilities of gels and liquid crystals and the interaction between sterols and phospholipids.

Our main concern here is with cell membrane permeability since it has direct bearing on sterol absorption from the intestinal lumen in man. In order to affect the structure and function of organelles, the sterols have to traverse the cell membrane. This membrane consists of free sterols, sterol esters, sterol glucosides, glycoproteins, glycolipids, glycotriglycerides, and a wide range of phospholipids, all in contact with proteins, carbohydrates, electrolytes, and water, and all interacting, according to *Rouser et al.* [710].

From *Smith* [776, 777] we learned that pleuropneumonia-like organisms require a specific configuration of sterols, namely a planar molecular with the radical 3-OH in equatorial position. Such a position resists hydrolysis, whereas cis-fused A/B ring esters of the axial 3-OH radical are easily hydrolyzed.

Grunwald [348] studied the permeability of red beet tissue. Alcohol increased the permeability for β-cyanin; methanol required 10–15 h incubation. Its effect could be inhibited by $CaCl_2$ and, even more effectively, by stigmasterol, β-sitosterol and, still more, by cholesterol. In contrast, ergosterol enhanced the methanol-induced leakage. The effect of sterols depends on their stereochemical structure since it is due to sterol-phospholipid interaction and charge distribution. Barley roots, *Hordeum vulgaris* var. Barsky, were grown in darkness at 21 °C for 3 days. The three main plant sterols and cholesterol were then added in various concentrations. Only cholesterol reduced leakage considerably. *Grunwald's* [349] findings concerning the inhibitory role of side-chain methyl, and even more so of ethyl, groups have been confirmed by other investigators. These findings, if quantitative cholesterol/phospholipid relationships are added as further consideration, aid in the explanation of sterol absorption in man.

b) Steroid Hormones in Plants

Phytosterols can act (a) as steroid hormones – phytohormones, and (b) as precursors of insect molting hormones – ecdysones. Phytosterols are either (a) steroid hormones as known to us from the animal world, or (b) precursors of cardenolides.

Heftmann [363] cites the legend of the estrogen-containing fruits of the pomegranate tree, the 'Tree of Knowledge', and the Moslems' myth that among the three things Adam took with him from the 'Garden of Eden' were estrogen-rich dates. He recalls the Egyptians' use of extracts from pomegranate seeds to induce in castrated mice the typical vaginal cornification of the Allen-Doisy test. This could be interpreted either as an effect of estrogen in the seeds or as an effect of β-sitosterol functioning as phytoestrogen. A combined effect presents a third possibility.

Heftmann [363, 368] reviewed the topic of steroid hormones in plants in 1967 and 1975. He listed a large number of plants in which steroid hormones were found, both estrogens and progesterone.

According to *Bennett and Heftmann* [68] and *Bennett et al.* [70], *Digitalis lanata* converted β-sitosterol-3-^{14}C to progesterone at a rate of 0.053%, to digotoxin at a rate of 0.24%, to ditoxigenin at a rate of 0.11%, and to digoxigenin at a rate of 0.086%. These amounts seem small; however, cholesterol-4-^{14}C was converted to progesterone at a rate of only 0.01% and to cardenolides, at a rate of 0.2%.

Estrogen and progesterone are not the only steroid hormones detected in plants. Desoxycorticosterone has been found in rice bran oil. This particular discovery is of interest since a link has been established between β-sitosterol and cortisol.

Steroid hormones are active within the plants proper. They regulate cell differentiation which expresses itself in flowering and in the sex characteristics of plants. In short-day and long-day experiments, the activation of the florigens, the flower hormones, is under photoperiodic control. Here again is seen an analogy between periodicity of florigens, biliary secretion, and intestinal coenzyme activity in man.

Heftmann [361] cites multiple experiments in which the addition of estradiol or testosterone to water, to a liquid culture medium, or to the soil influenced sexual differentiation of plants and/or increased either the number of female or of male flowers.

c) Phytosterols as Vitamins

There is an overlap, both biochemically and functionally, between steroid vitamins and steroid hormones. Strong efforts are under way to move the whole vitamin D complex from the roster of vitamins to that of hormones. While this is going on, plant sterols which give rise tro insect molting hormones have been called 'vitamins for insects'.

Ergosterol is considered the 'nucleus' of all D vitamins, ever since its physiologic role became known. In spite of efforts to reclassify the D complex, some phytosterols were added to the D vitamins not long ago by *Thiers et al.* [861, 862] who tried to clarify the intricate nomenclature of D vitamins: ergosterol is also called ergocalciferol, pro-ergocalciferol, vitamin D, vitamin D_2, pro-vitamin D_2, or ex-vitamin D_2. Vitamin D_3 is 5,6-cis-cholecalciferol; pro-vitamin D_3 is 7-dehydrosterol. Vitamin D_4 is 22:23-dehydro-5,6-cis-ergocalciferol. Vitamin D_6 is sitocalciferol. Ex-vitamin D_6 is stigmacalciferol. Campecalciferol is a vitamin D activator. Ergosterol, though acting as pro-ergocalciferol, lacks one of the vitamin's main properties, namely the antirachitic effect in children. *Lederer* [528] stated that ergocalciferol is less active than cholecalciferol because it is poorly absorbed by the intestinal mucosa.

In 1950, *Katsui* [447], under the heading of 'resources of vitamin D', mentioned that the sediment of saké yields 0.09% sitosterol. One may argue with the designation of sitosterol as a D vitamin but not with its presence in rice wine. It originates either from the penicillium mold in the sediment or from the barrel wood. *Braus et al.* [143] isolated sitosterol-β-*D*-glucoside as a factor in the clarity of whiskey. This substance was identical with the extract from oak wood. The bark of oak, *Fagus silvatica* L, contains betulin and free β-sitosterol, according to *Ludwiczak and Szczawiñska* [552]. *Matsui et al.* [559] isolated from 11 kg of the alkaloid-free fraction of opium 4.5 g β-sitosteryl-*D*-glucoside and from this, 0.9 g β-sitosterol.

Referring to its medicinal uses, *Paye* [644] called stigmasterol 'a new fat-soluble vitamin'. *Thiers* [859, 860], the first to refer to stigmasterol as a vitamin, identified it as 'antistiffness factor', essential for the guinea pig. Later, he broadened his concept, ascribing a wide range of therapeutic properties to the 'unsaponifiables' of vegetable oils.

d) Phytosterols as Source for the Manufacture of Vitamins and Hormones

The feasibility of using sitosterol as starting material for manufacture of vitamin D was investigated by *Dobrowsky and Kohl* [231]. They found this to be uneconomical since only 5 g of raw sitosterol could be obtained from 1 kg of pressed grapes. Crude sitosterol exposed to UV radiation, and 7-dehydro-cholesterol exposed to Mg spark radiation, yielded vitamin D_2 of a potency equal to 30 million IU/g. Prolonged radiation led to physiologically inactive 'suprasterol'.

Phytosterols can also be used as a source of steroid hormones. The technic of isolating sitosterol from wood in the course of paper manufacturing and the technic of converting β-sitosterol to testosterone, 17-methyltestosterone, and to other hormones has been thoroughly described by *Khaletskii* [464, 465]. He obtained 10–20 g of steroid hormones from 5,000 g β-sitosterol and found it a more economical starting material than cholesterol. The transformation of β-sitosterol to steroid hormones has been confirmed by *Subbiah* [804].

e) Phytosterols as Phytohormones

Phytohormones are of interest since their activity mimics and/or supplements the activity of steroid hormones present in plants. Either of the two, phytohormones or steroid hormones, may be too small to have significance but a synergistic effect could have importance. Let us recall – without citing references – that estrogens have been considered by some in the prevention of atherogenesis in chicken and as therapeutic agents for man.

Phytoestrogens. Between 1961 and 1974, *Elghamry* [247, 248], alone or with associates, published 33 essays on phytoestrogens, especially on β-sitosterol. The main sources of β-sitosterol for their experiments were *Trifolium repens* (Ladino clover), and an extract of Egyptian *Glycyrrhiza glabra* (licorice). The studies involved the endo- and myometrium, the uptake of ^{131}I by the thyroid gland of ovarectomized mice, and the contractions of pregnant and non-pregnant uteri. *Zayed et al.* [952] monitored the uterine weight of immature mice after β-sitosterol administration.

Yamashita [950] injected rabbits s.c. daily for 5 days with 0.2, 10, or 20 mg of ovarian sterol and steroid hormones. These had significant pro-gestational activity – as determined by increased carbonic anhydrase activity in the endometrium which had been removed 24 h after the last injection. Sterols were less active than steroids. Expressed in enzyme units per gram of tissue, the activity of methyltestosterone, androstenedione, androsterolene, stigmasterol, and β-sitosterol were, respectively, 395, 125, 150, 100, and 89.

Phytosterols and Adrenal Cortical Hormones; Cortisol. Therapeutic doses of β-sitosterol were given to 13 patients for 10 days by *Ristelhueber and Contesse* [693]. There was no inhibitory effect of adrenal gland secre-tion, as judged by the amounts of 17-keto- and 17-hydroxycorticosteroids in the urine before and after ACTH stimulation (2.5 mg/8 h i.v.).

Werbin et al. [914–917] fed male guinea pigs a diet containing ^{3}H-β-sitosterol for 1 week. Cortisol of high specific activity was isolated from a 7-day pooled urine sample, indicating conversion of β-sitosterol to cor-tisol.

f) Phytosterols as Source of Ecdysones

Ecdysones, as all the molting hormones are called, promote: (a) yolk deposition in the egg; (b) larval development of nymphae; (c) conversion of nymphae to larvae; (d) larval growth, and (e) conversion of larvae to pupae, where these hormones are stored. Up to this point, ecdysones could also be labeled as growth hormones. It seems doubtful, however, that their function stops with larval development.

Ecdysones are steroid hormones and, as such, they may well have the same function in insects as steroid sex hormones have in higher species. Whereas *Homo sapiens* – in this context possibly more appropriately called *Homo eroticus* – engages in sexual activity for most of life, other vertebrates and invertebrates, insects, and flowers and trees have well-defined mating seasons. The field of endocrinology does not lend itself to extrapolation from flora to fauna, or from one species to another. Yet, all sexual behavior is hormone-dependent.

Before proceeding to the molting hormones a brief mention should be made of the presence of other steroid hormones in insects. The only reference is that of *Marker and Shabica* [557], who extracted cantharidin

from the genital glands of the Spanish fly, *Cantharides Russian,* and found
a mixture of androstane and pregnane, with β-sitosterol as the principal
sterol. They found no estrogenic activity.

Much has been added to our knowledge of steroid insect molting
hormones since the topic was reviewed by *Heftmann* [365] in 1970. He
reported on the wide distribution of molting hormones in plants, and on
phytosterols as precursors of the ecdysones. He found that *Polocarpus
elata* seedlings converted cholesterol-4-^{14}C to ^{14}C-ecdysone and he discovered that *Calliphora erythrocephalus* larvae converted ^{3}H-cholesterol
to ecdysones. Thus, he established cholesterol as the starting material for
ecdysones, both in plants and in insects. Zoophagous insects obtain cholesterol from the diet; some of it is of endogenous origin. Phytophagous
insects obtain some cholesterol from plants but most of it is synthesized
from phytosterols.

The best-known ecdysones are ecdysterone and ecdysone. Some others are polypodine B, ponasterone A, B, C, and D, pterosterone, inokos-
terone, and cyasterone. All are related to cholesterol and desmosterol.

The activity of the ecdysones is expressed in Calliphora units: It
ranges from 5 U for ecdysterone up to 200 U for ponasterone D. The way
in which these hormones operate is not quite clear. They may activate
specific genes in chromosomes and produce the ribonucleic acids which
govern enzyme synthesis.

In animals, desmosterol represents the last major link in the chain of
cholesterol biosynthesis. The next are 5,7,22,24-cholestatrien-3β-ol and
7-dehydrocholesterol. In plants, the pathways are different – up to desmosterol – because of differences in α- and β-ethyl, methyl, ethylidene, or
methylidene groups, and the absence, presence, and location of double
bonds in side chains.

g) Inhibition of Molting Hormone Synthesis

Much has been learned about molting hormones through experiments
with 22,25-diazacholesterol. *Svoboda and Robbins* [825–827] and *Svoboda et al.* [824] established that desmosterol is the common intermediate
in conversion of C_{28} and C_{29} phytosterols to cholesterol. They experimented with several species but their results with *Manduca sexta* are the
most illustrative (table X).

Earle et al. [244] found that azasterols had no sparing effect on cholesterol and that they retarded larval development of the adult boll weevil,
Anthonomus grandis, although the effect on weight, on days of develop-

Table X. Influence of dietary azasterol addition on levels of desmosterol and cholesterol (as % of total sterols) in *Manduca sexta* fed various sterols. After *Svoboda and Robbins* [826]

Dietary sterols:	Desmosterol		Cholesterol	
22,25-Diazasterols	+	−	+	−
β-Sitosterol	21.8	0	5.9	84.5
Campesterol	14.4	0	7.7	74.5
Stigmasterol	12.2	0	5.1	75.0
Fucosterol	79.4	0	6.4	94.1
Chalinasterol	39.8	0	8.0	87.5
Brassicasterol	10.5	0	3.5	70.5
Dihydrobrassicasterol	11.3	0	2.8	74.4

ment, and on the yield of larvae differed widely with different dietary sterols.

One can conclude that the interference of antimetabolites – which, as mentioned before, can be used in biologic pest control – depends on four factors: (1) there are qualitative differences between the antimetabolites; (2) there is a quantitative relationship between antimetabolites and sterols; (3) the interference with the metabolic pathways differs from one phytosterol to another, and (4) the susceptibility or resistance varies from one insect family to another.

Part II

A. Dietary Supplements

Years before phytosterols or β-sitosterol or vegetable oils became known as cholesterol-lowering agents, attempts had been made to influence the plasma cholesterol level by modifying customary diets. Diets were supplemented by certain fruits or vegetables. Four such additives were used more often than any others: (a) the globe artichoke, *Cynara scolymus* (or *scolimus);* (b) the egg plant, *Solanum melongena,* L.; (c) the avocado pear, *Persea americana,* and (d) alfalfa, *Medicago sativa.* Some of these had been used as folk medicines, though not with the specific intention of lowering plasma cholesterol.

Artichoke. In the years 1933–1935, a group of French scientists presented a series of reports on the usefulness of 'cyneratherapy'. We will cite only a few of their papers. *Tixier and Eck* [870] injected 50–150 mg of the 'active principle' of *Cynara* ('artichaut') into rabbits and man. In man, blood cholesterol decreased from 523 to 277 mg/dl by the eighth day of treatment! A drawing of the X-ray film of the abdominal aorta of 1 patient illustrates the paling of the atheroma which allegedly accompanied the clinical amelioration [871]! The results of this happy ('heureux') effect of artichoke leaves were: enhancement of biliary secretion, stimulation of cholesterol metabolism and uric acid metabolism, and diuresis. *De Sèze* [220] called the extract of artichoke an aperitive, a choleretic and a cholesterolytic agent. His patients received either 2 g orally or 5 ml of a 2% solution intravenously for 2 days, and then 1–3 g orally for 7 days. He, and also *Eck and Desbordes* [245], claimed that rabbits injected with 1 mg adrenaline developed endogenous hypercholesterolemia and that this responded to intramuscular injections of the crystalline principle of *Cynara* for 14 days by a drop in blood cholesterol from 1,800 to 610 mg/dl.

Hermann [375] gave oral doses of 1 g of 'Acoucil', a powdered extract of *Cynara,* thrice daily to 10 patients with hypercholesterolemia. In 2 patients no effect was noted, in 1 patient serum cholesterol rose, and in 7, it decreased by an average of 24%. These results were better than those obtained by *Hermann* [375] using eggplant powder. *Mancini et al.* [556] successfully treated 23 patients by giving them 250-mg tablets of the 'active principle', i.e. quinic acid-1,4-dicaffeate (cynarine). All these reports sounded too good to be true. *Pomeranze and Chessin* [669] gave 'Chophytol', a preparation obtained from the Jerusalem artichoke to 12 patients for 6 weeks without effect on serum cholesterol. (The dose was not disclosed.) They cautioned that the response to this and other hypocholesterolemic agents is not uniform.

Two reports, by *Atherinos et al.* [36] and by *Soliman* [782], are of particular interest: They obtained from 2 kg of defatted artichoke (Egyptian variety) 2.8 g of a neutral fraction and 10 g of an acidic resin. The latter contained 2 g of crystalline sitosterol, some stigmasterol, taraxasterol, ψ-taraxasterol, and cynarogenin, a trihydroxy-sapogenin.

Eggplant. Roffo [701, 702], in 1944–1946, was the first to use, as a 'decholesterolizing' agent, a 20% alcoholic extract of 'berenja' for s.c. injections into rabbits and man. At the start of the investigation healthy rabbits had an average serum cholesterol of 59 mg/dl. After 18 weeks of treatment, the average was 33 mg/dl. In man, such injections reportedly spurred the conversion of cholesterol to cholestanone, allocholesterol, epicholesterol, cholic acid, deoxycholic acid and lithocholic acid. The cholagogic and diuretic effects seemed more impressive than the hypocholesterolemic action.

The shapes and names of eggplants may be of some interest: the Chinese type, *Serpentinum Bailey,* is cylindrical. The American variety, *Solanum esculentum* Nees, is either ovoid/white or round/brown. The French have their 'aubergine', the Italians 'melanzana', the Spanish 'berenja', the Germans 'Eierkartoffel' or 'Eierpflanze'. More important than names and shapes, however, is the fact that these plants differ chemically, since this may explain the different results obtained in clinical trials. For example, the Dutch Oriental Indian plant contains 1% fat, the Indian Oriental Dutch, 2.1%, but the Indian Occidental Dutch, only 0.3%.

Hermann [375] gave 1 g of 'berenjana' powder orally three times a day to 4 patients with hypercholesterolemia. In 2 patients, the effect was slight, while in the other 2 it was moderate, i.e. total cholesterol (TC) decreased by 11%, and esterified cholesterol (EC) by 8%. *Pomeranze and Chessin* [669] had negative results with eggplant powder in tablet form. *Wilkinson*

[928] obtained inconclusive results with dessicated eggplant and many other medications. *Wilkinson et al.* [933] gave this preparation to 6 men, in 12- to 24-gram daily doses. Their negative results may have been due to the fact that 4 subjects had normal blood cholesterol levels while 2 had essential hypercholesterolemia.

Graham et al. [333] found eggplant extract to be somewhat hypocholesterolemic for rabbits but that it had no anti-atherogenic effect. On the other hand, *Mitschek* [598] found eggplant to reduce atherosclerosis in cholesterol-fed rabbits. According to *Kritchevsky et al.* [506] eggplant powder reduced cholesterol absorption in rats but was not hypocholesterolemic.

Avocado. Grant [337, 338] prescribed half an avocado pear or one and a half pears each day for 16 men. This amount corresponded to about 40 g of fat which had been eliminated from the diet. 8 men, of whom 3 had diabetes and 1 had hypercholesterolemia, failed to respond. In the other 8 serum cholesterol decreased by 9–53%, especially in the esterified portion, during a 9- to 85-day regimen. The effect was ascribed to either the unsaturated fatty acid (UFA) content (iodine number 94), or to the sitosterol content of the pears (ca. 0.8 g/day), or to a combined effect of the two.

Avocado oil has 2–4% of unsaponifiable matter; *Paquet and Tassel* [643] analyzed this fraction by GC, CC, and TLC. They found in the California variety 45% sterols and, in the Israeli variety, 14% sterols. Among the sterols the California type contained 80% β-sitosterol and 6% campesterol. They claimed to have found traces of cholesterol which is patently impossible in vegetable matter. The iodine number of the California pear was 162, that of the Israeli pear 220. *Prista and Alves* [676] extracted dried avocado leaves, used by natives of Angola as a diuretic, and found two flavons, quercitol and β-sitosterol. *Lamberton* [526] used the unsaponifiables of avocado and soja for treatment of scleroderma by Thier's method. The pharmacologic effect was ascribed to phytosterols.

Alfalfa. Cookson et al. [197] and *Cookson and Fedoroff* [198] first reported that dietary supplements of alfalfa inhibited cholesterol-induced atherosclerosis in rabbits. *Horlick et al.* [403] noted the hypocholesterolemic effect in rabbits that had been fed alfalfa and ascribed this to interference with cholesterol absorption. This may be due to the bulk effect of undigested fiber, the plant sterols of alfalfa complexing with cholesterol, or to unknown factors. When used to replace cellulose in a semipurified,

cholesterol-free, atherogenic diet, alfalfa significantly reduced the atherogenicity of casein, according to *Kritchevsky et al.* [503].

Cole and Krusberg [183] pointed out the high content of α-spinasterol
and Δ⁷-stigmastene-3β-ol in alfalfa callus tissue. *Deuel* [224] lists alfalfa as
a major source of spinasterol and alfalfa seeds as a source of γ-spinas-
terol.

Although the value of these dietary supplements may be debatable
their hypocholesterolemic effect might be attributed to their phytosterol
content.

B. Vegetarian and Lacto-Ovo-Vegetarian

Whereas fruits and vegetables are usually taken as adjuncts to diets
and not as exlusive nutrients, the aim of vegetarian diets is to replace
animal foodstuffs, including animal fats. Vegetarian diets are adhered to
for various reasons – they are not primarily prescribed to depress plasma
cholesterol. They are most often linked with religious dietary laws. Often,
they are fads.

The amount of animal fat in such diets depends on the compliance
with dietary laws. Individuals are not suited for studies. Comparisons have
to be made between large, mostly religious, groups. The lacto-ovo-vege-
tarian diet may be closer to that of non-vegetarians. Strict non-vegetarians
do not exist, though some omnivorous people consume few vegetables and
fruits.

Hardinge et al. [355] and *Hardinge and Stare* [356] studied the serum
cholesterol of 'young' (adolescent) and 'old' (adult) vegetarians. They
found a significant positive correlation between percent of animal fat
(hexadecenoic acids) and serum cholesterol in the adults but not in adolescents or in pregnant women. This correlation applied to the subjects
regardless of diet. They also found a significant negative correlation
between the intake of UFA, especially PUFA, of linoleic acid and oleic
acid versus that of cholesterol – again, in adults only. They tabulated the
total caloric intake and the calories obtained from fat, but ignored those
from protein. Recently, it has been suggested that the low cholesterol levels
found in vegetarians may be due to dietary fiber.

It is debatable which is more relevant to blood cholesterol levels, total
fat ingestion or the amount of saturated fatty acids. Two studies, both on
large populations, are cited here for consideration. *Shaper* [757] reported
on the low serum cholesterol of the Samburu tribe of Kenya who subsist on

a milk diet. *Pollak* [665] described the low incidence of atherosclerosis in the Thai, whose diet of 1,209 cal/day comprised 26% fat, the fat being coconut oil (with 91% saturated fatty acids – SFA, 2% linoleic acid, iodine number 9–10) and lard (with 43% SFA, 10% linoleic acid, iodine number 40–60).

West and Hayes [918] studied a group of Seventh Day Adventists. Of 3,260 church members, 1,724 volunteered for the study. Of these, 705 were vegetarians, 1,019 were not. The mean cholesterol was 196 in the former group and 185 mg/dl in the latter. This difference of 11 mg was statistically significant at p < 0.01. Curiously, the difference was not found in the 15- to 24-year olds. *McCullagh and Lewis* [566] studied 44 Trappist monks aged 40 years and over. The subjects were lacto-vegetarians; 26% of their caloric intake came from fat, with one half of it being butterfat. Their blood cholesterol levels were lower than those of a comparable group of men who ate an ordinary diet. Yet, 26 of the 44 monks (59%) had clinical atherosclerosis and 21 (48%) had hypertension. *Chen* [166] examined 145 nuns and priests in Chinese Buddhist temples – where absolute vegetarianism prevailed for centuries – and compared them with 265 other adults. The respective blood cholesterol values were 156 and 176, triglycerides (TGs) 86 and 103, and phospholipids 493 and 527 mg/dl.

Whitney et al. [919] determined campe-, stigma-, and β-sitosteryl acetates in the serum of vegetarians and non-vegetarians. In 5 male and 5 female vegetarians the three respective phytosteryl values were 7.4, 7.3, and 9.5 n/ml. In 9 non-vegetarian males the three values were 27.8, 29.6, and 17.6 n/ml, and in 9 non-vegetarian females they were 22.2, 23.3, and 10.1 n/ml. The differences between the sexes in non-vegetarians could not be explained. The report lacks data on total caloric intake and also on the composition of the diets.

In a study of 48 infants with kwashiorkor, *Schendel and Hansen* [723] state: 'Pre-kwashiorkor diet is an almost total vegetable diet. One of the sterols in such diet, sitosterol, has been associated with reduced absorption of cholesterol – and hypocholesterolemia is one of the features of kwashiorkor.'

There seems little doubt that a strict vegetarian diet leads to lower plasma cholesterol values, especially in middle-aged subjects. The lack of response to such diet by adolescents has not been explained. It is regrettable that none of the reports on vegetarians give information on the intestinal microflora which surely must undergo drastic changes under the influence of vegetarian diets.

C. Terminology of Dietary Fats

It has already been stated that it is not feasible to separate sterols into zoo- and phytosterols. It is possible, however, to speak of animal fats and of vegetable or plant fats – as long as one refers to their origin.

FAs can be separated into saturated and unsaturated types. The latter are subdivided into mono- and polyunsaturated acids or, to use the proper terminology, into mono- and polyenoic FAs. To speak of saturated and unsaturated fats is incorrect since all fats contain multiple acids of varying degree of saturation or unsaturation. The more UFAs, the higher the iodine number; the more SFAs, the lower the iodine number of a fat. Though linoleic and arachidonic acids are representative of unsaturation there are some highly unsaturated oils which lack or almost lack these two acids. It is misleading to place animal and vegetable oils into juxtaposition, as though all animal fat were saturated and all vegetable fat unsaturated. Coconut oil, a plant oil, has a very low iodine number. To speak of animal fat as 'hard' and of vegetable oils as 'soft' is also incorrect. Solid fats can liquefy and liquids can solidify. Butter and butter oil or cocoa butter and cocoa oil are representative examples.

In 1930, *Gardner and Gainsborough* [293] lamented the fact that substances with no chemical resemblance, such as cholesterol, lecithin, and neutral fats, are often lumped together under the misleading heading of 'lipoids'. Fats are lipids; lipid-like substances are lipoids. *Mead* [569] criticized the misuse of terminology and pointed out that 'vegetable fats' contain largely oleate glycerides, that 'seed fats' and marine animal fats contain largely PUFAs, and that the so-called essential fatty acids (EFAs) may not be essential at all.

The confusion of terminology may be best illustrated by reference to an essay written by one of the most renowned nutritionists of his time, *Kinsell* [472], in which he used the terms 'vegetable fats', 'vegetable oils', 'vegetable sterols', and 'vegetable phospholipids' as if they were interchangeable. In view of the uncertainty concerning the nature of his starting material for dietary studies it is not surprising that *Kinsell* and co-workers alternatingly ascribed the cholesterol-lowering effect of vegetable fats and oils – a concept not universally shared – to lecithin-cephalin [474], to phospholipids [469, 473, 475, 477], to soy sterols [478], to 'some material' similar to stigmasterol [475], to stigmasterol *per se* or to β-sitosterol [469], to EFAs [480], to linoleic acid [476, 482], or to USFAs [476].

D. Vegetable Oils as Cholesterol Depressants

Vegetable oils, soy sterols and β-sitosterol were all suggested as hypo-cholesterolemic agents at about the same time. The first research projects were initiated almost simultaneously and the first publications regarding these agents were barely 1 year apart.

The bibliography on animal experiments and clinical trials is tremendous, and cannot be covered in this book. The gaps in the data, as indicated by the many questionmarks in our tables, are also discouraging. Because of lack of pertinent information, many essays can only be referred to lightly, by numbers pointing to the bibliography.

The first reports on the use of vegetable oils were based on clinical observations. Animal experiments followed. Therefore, we will survey the clinical trials first, experimental results next. A discussion of the theories concerning the active ingredient of vegetable oils is included.

1. Man

Multiple reports have been issued by many institutions and they are briefly summarized in this section. The number of subjects for whom animal fat was replaced by dietary vegetable oils varied from one study to another, anywhere from a single patient to 140. For presumably healthy persons, the number averaged 20 per study, for patients 28. Some subjects were called 'young', some were called 'middle age', a few were said to be 'old'. All these terms are vague. The sex of the subjects has rarely been mentioned. The duration of studies varied from 6 days to 6 years. Often, the length of the study has not been indicated. Six studies lasted less than 2 weeks, 13 lasted less than 6 weeks, one extended over 6 years. The average duration was 11 months. Information about the calories contributed by vegetable fats to the daily total is mostly missing. The dose of oil given varied from 15 to 180 g/day; the average dose being 100 g/day. Several authors reported that they used 'vegetable oils', others specified the use of corn oil or of safflower oil, and less often the use of soybean, peanut, rapeseed, rice, sunflower seed, sesame, or olive oil. Many compared the effect of several oils and also margarines. Data on the initial plasma cholesterol level are sparse. Where they were given, the levels were about the same in 'healthy' subjects and in those with atherosclerosis, 194–360 mg/dl, with an average of 262 mg/dl. Some clinicians used vegetable oils in an attempt to lower cholesterol level prior to starting a dietary regimen. About one third of all subjects responded to such vegetable oil

Table XI. Influence of vegetable oils on plasma cholesterol levels in man

Number of subjects	Oil fed	% Calories	Duration months	Change in cholesterol level, %	Ref.
18	'vegetable'	–	–	0	29
40	'vegetable'	–	40	–22	260
10	corn	–	3–9	0[1]	254, 255
9	corn	–	3	–10	690
6	corn	40	3–4	–20	191, 192
11	corn	40	3	–24	398–401
20	corn or soya	–	4	–20	750
–	safflower/coconut safflower/cottonseed or corn	–	–	0	819–823
10	safflower	20	1	–27	214
327	soybean	–	74	– 9	880–882

[1] Administered after 12 months on a low-fat diet.

regimen with a decrease in the plasma cholesterol level. Where such a decrease occurred it amounted to 1.3–32% of the original level. The mean decrease for those who responded was 16%. This reduction occurred regardless of the type of oil ingested, the dose taken, the length of the trials, or the condition of the subjects. The results are distorted by the manifold modifications in experimental design and by the fact that the degree of reduction of plasma cholesterol has been reported by some in milligrams and by others as percent drop, often without indication as to the starting level (table XI) [9–11, 18, 24, 106, 107, 109, 111, 113–115, 150, 151, 218, 291, 334, 335, 404, 483, 484, 489, 553–555, 637, 758, 769, 797, 839, 857, 872, 873, 934, 940, 942].

The first proponent to replace animal fat by vegetable fat was *Kinsell* and co-workers. They contributed 16 essays on this topic between 1952 and 1958 [469–484]. Their observations were confirmed by many, but were also opposed by many.

Several authors found a low-fat diet *per se* effective; others combined a low-fat diet with vegetable oil ingestion [321, 322, 388, 389, 398–401]. *Mayer et al.* [563] advocated a combination of a vegetable oil regimen with a low-fat, low-cholesterol diet for best results. *Gordon and Brock* [320] considered also the effect of carbohydrates, protein and cellulose during a vegetable oil regimen. A report by *Walker et al.* [902] on the decrease of serum cholesterol in young women who, for 6 weeks consumed either 50 g vegetable protein or 50 g animal protein per day (ca. 95 g fat/day), puzzled the referee of *Nutrition Reviews* [625].

2. Animals

A survey of animal experiments with vegetable oils illustrates the difference in results from those of clinical trials. The experiments are summarized in tabular form, with the species arranged in alphabetical order (table XII).

Hegsted [369] was the first to caution against extrapolation of results from animal experiments to man. The preceding table and subsequent discussion support his view.

Table XII. Influence of vegetable oils on serum cholesterol level of animals

Species	Oil		Duration	Dietary cholesterol	Results	Ref.
	type	level				
Chicken	Fish	10%	ND	+	D	941
	Fish	–	ND	+	D	589, 590
	CO, CSO	–	ND	+	I	
	CO	10%	36 months	–	O	300
	CO		6 months	+	D	275–278
	CO/CSO		ND	ND	D	370
	CSO	5%	ND	+	D	636
	CSO		24 months	+	D	653
	SBO/CNO		20 weeks	+	D	57
	SSO	10%	8 weeks	–/+	I/D	48
Dog	SFO		5 months	ND	D	233
	SO		10 days	+	D	342

CNO = Coconut oil; CO = corn oil; CSO = cottonseed oil; MO = mustard oil; OO = oat oil; PNO = peanut oil; RBO = rice bran oil; SBO = soyabean oil; SFO = safflower oil; SSO = sesame oil; SUFO = sunflower seed oil; LA = linoleic acid; ND = no data; D = decrease; I = increase; O = no effect.

Table XII. (continued)

Species	Oil		Duration	Dietary cholesterol	Results	Ref.
	type	level				
Pigeon	SFO	10%	12 months	+	D	173, 675
Minipig	SBO		48 weeks	+	O	390
Boar	CO		ND	–	I	142
Rat	CO	20%	ND	–	I	40
	CO		ND	–/+	D	756
	CO (7.5% LA)		8 weeks	+	O	419
	CO, OO (10%)		ND	+	D	674
	PNO (28% LA)		7 weeks	+	I	632
	RBO	10%	8 weeks	–	I	680
	SUFO	15–20%	ND	+	I	923, 925
Rabbit	CO	20%	28 weeks	–	D	280
	CO, SBO, OO		36 days	–	D	940
	SFO	20%	100 days	–	D	525
	SFO	25%	14 weeks	–	D	593
	CNO		ND	ND	I	579
	PNO		9 months	+	I	804
	CO	19%		+	I	32, 863
	CO, LSO		84 days	+	D	567
	MO, SSO		3 months	+	I	167
	PNO	6%	2 months	+	I	507
	SBO		96 h	+	D	898
	SFO	12%	125 days	+	O	798
	SUFO		ND	ND	O	341
Primates						
Cebus	CO	25%	15 weeks	+	D	937, 938
	CNO	25%	15 weeks	+	I	
Macaca	PNO	56% cal.	17 weeks	–	I	318
radiata	SSO	56% cal.	38 weeks	–	I	
	CSO (7.5% LA)		ND	–	D	419, 420
Rhesus	CO		8 weeks	+	I	670, 671
	CO	25%	ND	+	I	901
	PNO	25%	ND	+	I	
	SFO	20%	6 months	–	D	251
	SSO, MO, LNO	20%	ND	–/+	O/D	46, 47

CNO = Coconut oil; CO = corn oil; CSO = cottonseed oil; MO = mustard oil; OO = oat oil; PNO = peanut oil; RBO = rice bran oil; SBO = soyabean oil; SFO = safflower oil; SSO = sesame oil; SUFO = sunflower seed oil; LA = linoleic acid; ND = no data; D = decrease; I = increase; O = no effect.

All the reports on pigeons and dogs record a positive effect of vegetable oils, i.e. a reduction of serum cholesterol. Yet, the number of these reports is small. For pigs, one group of authors reported no change in the serum cholesterol level; in minipigs, another group reported a decrease on ingestion of corn oil in boars. For rats, four investigators reported a decrease in serum cholesterol, regardless of whether cholesterol was fed or not, while another author reported a rise in cholesterol. For primates, we have six positive and four negative reports, with the results unrelated to species, type of oil given, or length of the experiment.

Some essays deal with the influence of vegetable oil diets on the distribution of cholesterol in various organs, especially in the liver, rather than with the effect on plasma cholesterol. In rats fed with a low-fat diet and 2–10% corn oil for 3 weeks, organ cholesterol increased, according to *Gerson et al.* [296]. In mice fed with small amounts of sesame oil or linseed oil (0.5 g) plus large amounts of cholesterol (10, 20 or 50 g) there was no cholesterol retention in the body [725]. In rats fed cholesterol for 7 weeks, the subsequent intake of cottonseed oil increased serum and liver cholesterol [632]. Wheat germ, cottonseed oil or safflower oil reduced liver cholesterol of hypercholesterolemic rats within 9 days [17]. Results reflect the experimental design. *Russell et al.* [711] reported greater increases in liver cholesterol in normal and in alloxan-diabetic rats fed with 10 or 30% corn oil than in rats fed with 10 or 30% lard. Some [127, 189] believe that the effect of vegetable oils has a dubious value since it merely redistributes cholesterol in the body.

Some reports deal with the effect of diet on atherosclerosis. Since in chickens and rabbits hypercholesterolemia leads to atherosclerosis, inhibition of hypercholesterolemia prevents or lessens atherosclerosis. Reports on the favorable effect of vegetable oils in chickens outnumber the unfavorable by 3:1. This has not been explained, though in most studies cholesterol had been added to the rations. For rabbits, positive and negative reports are roughly 1:1 – but here reduction of serum cholesterol occurred when the rabbits were on a cholesterol-free diet and not when cholesterol was added. *Libert and Rogg-Effront* [542] observed significant aggravation of atherosclerosis in rabbits on a diet of 500 g carrots, 50 g bran and 2 g cholesterol in corn oil, though not as severe as on replacement of corn oil with peanut oil. *Kritchevsky et al.* [507] also found peanut oil most atherogenic for the rabbit, as did *Vesselinovitch et al.* [901] for the rhesus monkey.

The types of oils used for animal experiments were the same as those used in clinical trials. The degree of plasma cholesterol depression, where it occurred, was comparable in animals and humans. Animal experiments have contributed little to our knowledge.

3. The Active Principle in Vegetable Oils

Several theories have been advanced to pinpoint the ingredient responsible for the cholesterol-depressing effect of vegetable oils. Phosphatides, phytosterols – particularly β-sitosterol –, EFAs, linoleic acid alone, and UFAs in general (iodine number), were held responsible. None of these concepts has found universal acceptance.

a) Phosphatides

In a frantic search for an explanation of the apparent hypocholesterolemic effects of vegetable fats, as mentioned in the introduction to this chapter, *Kinsell et al.* [474, 475, 478] used a lecithin-cephalin mixture derived from soya as one of their several preparations, and they ascribed its effect to phosphatides. This concept was short-lived. At about the same time, the *Hormel Institute* [405] reported serum cholesterol depression in hypercholesterolemic rats fed wet-milled germ or soybean phosphatides.

b) Sitosterol

Beveridge and co-workers [104–121] were the first to suggest that phytosterols in the vegetable oils are to a large extent responsible for the hypocholesterolemic effect of oils. *Beveridge* [105] believed strongly enough in his concept to apply for a patent based on the role of β-sitosterol. 'Less than one half of the amount of sitosterol in corn oil has the same effect as corn oil as such when given as 25% of the total caloric intake.' He based this opinion, to which he still adheres, on numerous studies of large numbers of healthy persons who consumed various diets, including fractions of oils obtained by vacuum distillation. Maximum reduction of plasma cholesterol resulted from ingestion of the most volatile fraction, that which contained 9.95% of unsaponifiables. Its iodine value was somewhat lower than that of other fractions. *Nutrition Reviews* [622] commented that, in spite of the fact that corn oil has six times the amount of unsaponifiables than other oils, the amount supplied in clinical trials could not possibly suffice to effect cholesterol depression.

Many supported the 'sitosterol concept'. Among them were *Connor et al.* [192], *Moses* [597], and *Roels and Hashim* [700]. *Ahrens et al.* [12, 13] lowered serum cholesterol in a group of men by 20%, by feeding them a totally synthetic diet with 20% corn oil (iodine number 125; unsaponifiables 1.9%). On subsequent ingestion of sitosterol, an additional drop of the serum cholesterol by 9% was obtained. Reports by *Bronte-Stewart et al.* [147] and by *Kinsell et al.* [482] caused *Keys et al.* [460–463] to revise their 'predictive formula' for calculation of the effect of vegetable oils. At first, this formula was based on the SFA/UFA ratio; now, they recommended a correction to account for the phytosterol effect.

Many disagreed: *Jones et al.* [429], working with chicks, *Potter et al.* [674], working with rats, *Malmros* [553], and *Farquhar and Sokolow* [264], studying men, pointed out the small amounts of phytosterols in the oils. *Wood et al.* [942] found that PUFA alone, in absence of sterols, lowered serum cholesterol of men by 20% in 10 days. *Gordon* and co-workers [320–323] doubted that non-saponifiables were responsible for the effect of oils since 3 patients did not respond to the unsaponifiable fraction of pilchard oil and safflower oil whereas they did respond to these oils as such. *Avigan and Steinberg* [40] found that menhaden oil, with only 0.2% sterols, had the same effect as corn oil with 1.5% sterol content. *Kinsell* [473] commented that phytosterols are effective only when given in amounts exceeding the quantity in which they are present in natural fats, although in earlier reports [480–483] he had stated that phytosterols are responsible for the cholesterol-lowering effect of vegetable oils. The referee of *Nutrition Reviews* [621, 623] discounted the role of β-sitosterol and favored the role of PUFA in the oils. The question was raised [618] regarding the estimated caloric intake from animal fat (1,800 mg) and that from vegetable fats (23 mg). *Lovelock* [549] proclaimed in a discussion that: 'At least, *Kinsell* killed the red herring, i.e. that vegetable oils lower serum cholesterol because of their plant sterol content.' *Kinsell* had never made such a statement and, to this date, it has not been proven that β-sitosterol plays no role, or a lesser role than other ingredients, in vegetable oils.

The argument revolves around the amount of phytosterols. *Lown et al.* [550, 551] suggested that vegetable oils contain 1–2 g of plant sterols when supplied as 40% of the caloric intake. *Miller et al.* [592] and *Lambert et al.* [525] give the figures for unsaponifiables as 0.4% for safflower oil and 0.13% for hydrogenated coconut oil. An editorial [571] refers to 2% unsaponifiables in corn oil, 0.3% in wheat oil, and 0.2% in cottonseed oil.

Table XIII. Phytosterols (%) in four vegetable oils

Corn	Cottonseed	Olive	Soybean	References
1.6	0.9	1.2	0.8	195
1.3	0.6	0.8	0.6	415
0.6	1.0	1.6	1.9	352

Anderson and Moore [29] found 2.01% of non-saponifiables in crude corn oil, and 1.68% in edible refined corn oil. *Said et al.* [713] reported an identical hypocholesterolemic effect for corn oil and for lettuce salad oil with 2.8% unsaponifiable matter. *Beveridge et al.* [112, 117] found 0.5% unsaponifiables in corn oil, 0.3% in safflower oil. *Auterhoff and Nickoleit* [39] estimated total sterols in three samples of wheat oil, as 2.6, 2.62, and 2.68%. The figures vary with the source of the starting material and with methodology, both with regard to total sterols and to partition between several phytosterols (table XIII).

The statements concerning the small amount of sterols fade in the light of the large doses of oils prescribed for patients. The concept that phytosterols are largely responsible for the cholesterol-lowering effect of vegetable oils recently (1976) received support from *Borgström* [138] who made this the leading paragraph in the summary of a chapter on phytosterols.

c) Essential Fatty Acids

Two FAs have been singled out as essential for life: Arachidonic acid, $\Delta^{5, 8, 11, 14}$ -eicosatetraenoic acid, *in vivo* is converted to linoleic acid, $\Delta^{9, 12}$ -octadecadienoic acid. Man requires five times as much arachidonic acid as the rat, according to *Sinclair* [772]. *Mead* [569], discussing a report by *Kinsell et al.* [480] in which EFAs were emphasized, stated that linoleic acid is not essential and certainly cannot prevent atherosclerosis.

The leading voice in favor of the role of linoleic acid was once again that of *Kinsell and Michaels* [476] and *Kinsell et al.* [482]. They admitted that large doses of linoleic acid are required to obtain cholesterol depression. In a panel discussion [931], *Kinsell* defended his view but *Stare* and *Keys* pointed out that safflower oil (iodine number 140) with 75% linoleic acid is far more unsaturated than corn oil (iodine number 140), and that sardine oil (iodine number 186) is even more unsaturated though it lacks linoleic acid. A switch from safflower oil to sardine oil in the diet causes a

Table XIV. Iodine values and oleic acid (18:1) and linoleic acid (18:2) content (%) of various oils. After *Fieser and Fieser* [273]

Non-drying oil	Iodine number	18:1	18:2	Drying	Iodine number	18:1	18:2
Coconut	8–10	6.0	2.5	Cottonseed	103–111	33.0	43.5
Palm kernel	15–18	16.0	10.0	Corn	117–130	46.3	42.0
Palm	51–58	43.0	9.5	Soybean	124–133	33.6	52.0
Olive	80–85	83.0	7.0	Tung	100–180	15.0	0
Peanut	85–90	60.0	21.0	Linseed	170–185	5.0	61.5
Rape	94–106	29.0	15.0				
Beef tallow	32–47	48.3	2.7				
Lard	46–66	41.2	5.7				

rise, rather than a fall in serum cholesterol. *Anderson et al.* [25–28] found that safflower oil with a 23-gram linoleic acid content depressed serum cholesterol by almost 10%, the same oil with only 3 g linoleic acid, by only 6%. *Schroeder* [750] thought that linoleic acid is responsible for the effect of soy bean oil. In 20 men, 'an intake of 0.2–0.5 g linoleic acid per day as corn or soybean oil' was effective. Those who criticize the small amount of sitosterol in oils may also criticize the small amount of linoleic acid in some effective oils.

Bronte-Stewart et al. [147, 148] obtained the same degree of serum cholesterol reduction whether they fed the UFA fraction of pilchard oil (which lacks linoleic acid) or safflower oil (with 59.8% linoleic acid). They felt that sitosterol competes with cholesterol for solubility in oils. *Avigan and Steinberg* [41] gave menhaden oil, with little linoleic acid or arachidonic acid but an iodine number of 126, as sole dietary fat and had better results than with corn oil (42% linoleic acid, iodine number 178). *Ahrens et al.* [14] confirmed these results. *Suzuki et al.* [821–823] found no correlation between the effect of six margarines and their linoleic acid content. Rice oil, with less linoleic acid, was more effective than oils with a higher content. Whereas *Horlick and Craig* [404] reported identical results with corn oil and with free ethyl ester of linoleic acid or linoleate, *Stamler et al.* [787, 789], on the basis of experiments with chickens, concluded that the oleic acid in oils is more important than the linoleic acid content of the oils. The amounts of linoleic acid and oleic acid differ in various oils and the degree of unsaturation does not depend on the linoleic acid alone (table XIV).

Table XV. Iodine values of various oils and fats

Oil or fat	Iodine number
Coconut	2– 10
Hydrogenated	6
Cocoa butter	32
Lard	58– 66
Soybean	64–163
Partially hydrogenated	107
Hydrogenated	40
Safflower	75–140
Cottonseed	103–115
Hydrogenated	71– 93
Corn	108–130
Sardine	110–190
Sunflower seed	132–148
Hydrogenated	76
Codliver	150–160
Tung	190
Tuna	203
Menhaden	203

The data in the table and in the text are not fixed. Thus, one finds the amount of linoleic acid in sunflower oil, to mention a single oil, listed as low as 19% and as high as 72%. Again, the data depend on the source and on methodology.

d) Unsaturation

Since the iodine values, as a measure of the degree of unsaturation, were given in the preceding table, it seems desirable to summarize data collected from various texts (table XV).

The iodine number will differ with the method used for its assay. A report by *Horowitz and Winter* [406] on different iodine values of saf-flower *(Garthamus tinctorius)* oil from three generations of the same plant is of interest: in F_1 it was 91–101, in F_2 and F_3 it was 125–127.

As to the importance of unsaturation for the cholesterol-lowering effect of vegetable oils, opinion is divided. A case for PUFA was made by *Wood et al.* [942], after they had depressed serum cholesterol of man by 21% on a 10-day PUFA regimen. *Gordon et al.* [321, 322] supported this view on the basis of results with sunflower oil with the iodine number adjusted from 136 to 74 and peanut oil with the iodine number between 89 and 53. *Friskey et al.* [286] and also *Kinsell,* preferred to explain the hypocholesterolemic effect of oils on the basis of unsaturation.

Opposing the role of PUFA, *Kingsbury and Morgan* [468] found cod liver oil (iodine number 150, no linoleic acid) as effective as 25% corn oil (iodine number 115, 47% linoleic acid). *Hashim et al.* [357] and *Armstrong et al.* [33] obtained cholesterol reduction by 1.2–16.3%, unrelated to the iodine number of oils. *McCann et al.* [565] decreased the serum cholesterol by 16% in two thirds of their subjects fed peanut oil or safflower oil. *Seskind et al.* [756] could not explain the hypocholesteremic effect of vegetable oils in rats by the degree of unsaturation.

A third set of opinions is not quite as rigid as the 'pros and cons'. *Grundy and Ahrens* [344] pointed out the difference in response to PUFA between subjects with normal plasma cholesterol and those with various types of hypercholesterolemia. *Erickson et al.* [259] did not believe that dietary SFA or short-chain FA had a cholesterol-elevating effect in absence of dietary cholesterol. *Engelberg* [254, 255] was first to ask which of the two factors would be decisive – the reduction of SFA or the increase in UFA in the diet. Soon, *Kinsell* [471] declared that the degree of saturation of dietary fats is more important than their source, i.e. animal or vegetable origin.

Connor [187, 188] and *Connor et al.* [190, 191] stressed that, on a cholesterol-free diet, serum cholesterol decreases in spite of relatively saturated fats (iodine number 64) in the diet. They postulated that the plasma cholesterol decreased when the P:S (PUFA:SFA) ratio reached 1.5 or 1.6. Such a ratio corresponds to a combination of 52.5% olive oil plus 39.5% safflower oil or partially hydrogenated soybean oil. *Brown and Page* [151, 152] stated that the P:S ratio must be above 1.5 – which occurs when PUFA exceed 30% of total calories. Such a ratio would bring about a 6% decrease in serum cholesterol. *Gordon and Brock* [320] agreed with this formulation. *Hegsted* [369] postulated a P:S ratio of 2:1 since SFAs are twice as potent in increasing serum cholesterol as PUFAs are in decreasing the cholesterol level.

The controversy became further complicated. *Gordon et al.* [321, 322] claimed that PUFA not only lowers serum cholesterol but also causes an increase in acid and neutral fecal steroids. Thus, PUFA enhances the catabolism of cholesterol and its excretion, too. *Roels and Hashim* [700] claimed that the type of FA influences absorption, transportation, cholesterol synthesis and catabolism. *Lewis* [541] stressed only the increased catabolism. Increased bile acid secretion anteceded the decrease of serum cholesterol. *Connor* [188] and *Connor et al.* [190, 191] found that PUFA enhances bile secretion in normal persons and those with hyperlipemia V, but not in those with type II hyperlipemia.

4. Hydrogenation

The degree of unsaturation of FA depends on the number of double bonds. Monounsaturated acids (MUFAs) have one double bond, PUFAs have two, three, or more. Hydrogenation, the introduction of hydrogen, can be partial or complete and depends on the number of disrupted double bonds. The degree of hydrogenation is measured by comparing the iodine number of the end product with that of the starting material. Hydrogenation can be accomplished by chemical or physical means. Heating is one way to start the process. Of course, to those who do not hold PUFA responsible for the hypocholesterolemic affect of vegetable oils the question of hydrogenation is irrelevant.

A few animal experiments are cited to illustrate the diverse results obtained with different species. *Opdyke and Walther* [636] could lower the serum cholesterol of chicks by feeding them refined cottonseed oil but not by feeding them the crude oil. *Nath and Harper* [605] and *Nath et al.* [606–608] found that feeding rats with mixtures of 1, 2, or 25% hydrogenated coconut oil plus 1% corn oil led to a serum cholesterol value of 88 mg/dl and a liver cholesterol of 84.6 g. When cholesterol was added to the diet, serum cholesterol rose to 273 mg/dl and liver cholesterol to 3,070 g. *Diller et al.* [226–228] reported that in absence of dietary cholesterol the degree of unsaturation did not affect plasma or liver cholesterol of rats, however, on addition of cholesterol to the diet, that is to mixtures of non-hydrated and of completely hydrated corn oil or safflower oil, liver cholesterol rose and reflected the degree of unsaturation of these oils. *Jacobson et al.* [418] found in newborn calves the lowest serum cholesterol increment on a diet of hydrogenated soybean milk, the highest increase on a diet of crude, non-hydrogenated soy oil milk.

There are those who do not believe that hydrogenation affects the cholesterol-lowering ability of oils. *Wilcox and Galloway* [922] obtained the same degree of depression in men whether they ingested crude or refinded cottonseed oil. *Grasso et al.* [339] had similar results comparing natural and mildly hydrogenated soy oil. *Horlick* [399–401] acknowledged the formation of short-chain FA and of isomers of long-chain FA on heating of soybean oil but thought that heating corn oil to cooking temperatures for a short time had no influence on its efficacy as cholesterol depressant.

Others believe that hydrogenation decreases or nullifies the hypocholesterolemic effect of oils. *Gordon et al.* [321, 322] reduced the iodine number of sunflower seed oil from 130 to 70 and that of peanut oil from 87 to 75, by hydrogenation. This resulted in a lowering of the hypocholesterolemic effect of these oils. *Erickson et al.* [259] obtained a 23% rise in plasma cholesterol on a diet of partially hydrogenated soybean oil. *Malmros* [553] observed a 50% reduction of serum cholesterol with crude corn oil but only a 15% drop with hydrogenated corn oil. *Suzuki et al.* [819–823] stated that the more oils that are hydrogenated, the lower their hypocholesterolemic effect. *Bronte-Stewart et al.* [147] reported an increase in serum cholesterol on ingestion of hardened, hydrogenated coconut oil. *McOsker et al.* [568] found neither an increase nor a decrease of serum cholesterol on a diet containing partially hydrogenated cottonseed oil or soybean oil. Partial dehydrogenation caused the formation of up to 21% trans-FA and 8% trans-linoleic acid. *Tidwell and Gifford* [868] concluded from studies with various FAs that trans-isomers are less effective than cis-isomers in lowering serum lipids in humans.

A report by *Albanese et al.* [15] is important. For a period exceeding 5 years, 13 women substituted corn oil margarine (P:S=1.7) and peanut oil (P:S=1.5) with the conventional vegetable shortenings. Their serum cholesterol decreased significantly during the first 18–24 months, after which it began to rise until by 36–48 months it had reached the pretreatment level.

Nishida et al. [616] were the first to report that substitution of heated oil for fresh oil decreased the serum cholesterol of chicks but aggravated their atherosclerosis. *Kritchevsky and Tepper* [501] and *Kritchevsky et al.* [505] correlated the solubility of cholesterol in oils with the iodine number and with the number of carbons of FAs of various fats (table XVIa,b).

They confirmed increased atherogenicity of FAs. Heated corn oil was more atherogenic for rabbits than unheated corn oil. No difference was found between heated and unheated olive oil. The FA freed by heat-induced hydrogenation was held responsible for atherogenicity. A heated mixture of 6% corn oil and 2% cholesterol was more atherogenic than an unheated mixture. A heated mixture of 5.75% corn oil, 0.25% FA and 2% cholesterol was more atherogenic than the same mixture without FA. These reports provoked comments in *Nutrition Reviews* [629] which pointed out that the weight gain of rabbits fed heated oil, and the fact that

Table XVIa. Solubility of cholesterol (%) in various fats at 37°C. After *Kritchevsky and Tepper* [501]

Oil or fat	Iodine number	Solubility
Coconut oil	10	4.97
Butter oil	40	4.12
Lard	66	3.63
Corn oil	120	3.59
Codliver oil	160	2.70

Table XVIb. Influence of fatty acids (2%) on solubility of cholesterol in corn oil and coconut oil at 37°C. After *Kritchevsky and Tepper* [501]

Fatty acid	Corn oil	Coconut oil
Caproic (6:0)	4.47	5.83
Capric (10:0)	4.52	5.70
Palmitic (16:0)	4.03	5.59
Oleic (18:1)	3.99	5.44
Linoleic (18:2)	4.06	5.43
Linolenic (18:3)	4.05	5.43
None	3.59	4.97

FAs are not the only products of hydrolysis and that monoglycerides (MGs) and glycerol produced during heating of oils, may all be factors enhancing atherogenesis.

One may also recall past reports that feeding rabbits heated cholesterol in the form of eggpowder cakes or just heated eggs led to degrees of hypercholesterolemia which reflected the temperature to which cholesterol, or the eggs, were heated.

5. Margarines

Since butter is a major source of saturated FAs it seems logical to replace it with margarines made from non-hydrogenated vegetable oils. However, the opinions on the usefulness of such margarines in our diet are divided.

Beveridge and Connell [106, 107] reported that of eight commercial margarines used in dietary studies of healthy men, two led to hypercholesterolemia, although to a lesser degree than that induced by butter.

Horlick [399, 400] recorded hypercholesterolemia in men on diets containing corn oil margarine, hydrogenated corn oil margarine and commercial margarine. *Brown* [149], preparing a practical vegetable oil food pattern, commented that most brands of margarines and also of hydrogenated shortenings increase serum cholesterol and thus should not be part of our diet.

Adverse reports are matched by favorable comments. *Wilcox and Galloway* [922] recorded in 8 subjects a 12% decrease in serum cholesterol on a 15-day diet with margarine, i.e. to the same degree as achieved with cottonseed oil or hydrogenated cottonseed oil. On a corn oil diet, however, the drop was 22%. *Suzuki et al.* [822, 823] first used five margarines with 45% safflower oil plus 55% hydrogenated cottonseed or coconut oil. Some of these preparations had a hypocholesterolemic effect. Next, they used five margarines with 95, 87, or 70% safflower oil plus 5, 13, or 30% wholly hydrogenated palm kernel, or 95% safflower oil plus 5% wholly hydrogenated safflower oil. These had a cholesterol-depressing effect. *Swell et al.* [840] tested five corn oil margarines and 10 other margarines in 10-week studies. In those with normal serum cholesterol, the level decreased by 12%; in those with elevated serum cholesterol, it dropped by 21%. On termination of the study, serum cholesterol rose by 27% above the initial level.

6. Summary

The evidence that plasma cholesterol in man can be reduced by ingestion of vegetable oils in lieu of animal fat is fairly convincing. It is clear that this cannot be accomplished simply by replacing animal fats with vegetable oils on a 1:1 basis. The P:S ratio in vegetable oil diets must exceed 1.5:1 and should preferably be 2:1. It seems desirable to require a concurrent reduction of the total caloric intake and a restriction of dietary fat and cholesterol. Weight gain should be avoided.

Not all individuals respond alike to all vegetable oils. Thus, it is necessary to select oils and their dose by 'trial and error'. *Schröder* [750] obtained a 29% serum cholesterol reduction on a soybean oil diet but only a 20% drop on a corn oil one. Corn oil proved superior to cottonseed oil, the cholesterol decrease being 22 and 12%, respectively, according to *Wilcox and Galloway* [922]. Corn oil was more effective than safflower oil in decreasing cholesterol levels. According to *Tobian and Tuna* [872, 873], corn oil gave a 15% decrease whereas the decrease on safflower oil was only 9%. *Anderson et al.* [25, 26] found safflower oil, cottonseed oil, and sardine oil equally effective, and corn oil less effective. Olive oil proved as good as a low-fat diet. *Suzuki et al.* [819] obtained vastly different results with eight oils.

Constant monitoring of the effect of a vegetable oil diet is necessary, especially since some subjects seem to become accustomed to the regimen and their plasma cholesterol reverts to pretreatment level.

Ahrens et al. [11] treated 40 subjects for over 40 weeks with 350 g corn oil daily (in six divided doses) and found the results 'variable and unpredictable'. *McCann et al.* [565] recorded a decrease in serum cholesterol by less than 16% in 8 subjects, by 16% in 12 others ingesting peanut oil, and by 16% in 6 and by 10% in 3 subjects ingesting safflower oil. *Engelberg* [255] succeeded in lowering the serum cholesterol and LPs of 14 patients on a 1-year, low-fat diet. 7 of these patients responded to a 90 g/day vegetable oil diet for 3 months with an additional decrease in cholesterol, the 7 others did not.

Apparently, a vegetable oil regimen leads to a decrease in plasma cholesterol in about two thirds of the number of subjects. For these, the average decrease is 16%. It is greater in those with elevated levels than in the normocholesterolemic.

The dispute concerning the active principle in the oils cannot be easily resolved. There seems no doubt that neither sitosterol nor linoleic acid is indispensible for the hypocholesterolemic effect. Though linoleic acid is representative of UFA, it is not the sole determinant of unsaturation. In 1958, a panel of experts [931] ascribed a joint role to linoleic acid, arachidonic acid, and to β-sitosterol. The argument concerning the low amount of sitosterol in oils is not valid, especially when we learn that 350 g corn oil ingested daily supply (at a 1.6% β-sitosterol content) 5.6 g of this phytosterol per day.

Correlation of the degree of unsaturation and hypocholesterolemic effect is poor. It is doubtful whether collation of all available data would give a clue to the ideal iodine value. The breaking point of iodine number 129 has been suggested. This seems unrealistic; an iodine value of 100 corresponds to olive oil which is considered 'neutral' as regards cholesterol lowering. Considering liquid fats as 'suitably unsaturated' might be useful. This area of diet controversy has not been resolved and the decision as to what level of unsaturation is good rests on empirical data. The usefulness of vegetable margarines is limited, as most products are made with 'partially hydrogenated' oils – without a hint as to the degree of hydrogenation.

We learned from animal experiments that cis-isomers are more effective hypocholesterolemic agents than trans-isomers of fatty acids, including linoleic acid. Transisomerization occurs during hydrogenation, also during heating of oils. Additional important information gained from animal studies is found in the enhancement of atherogenesis by corn oil, especially by heated corn oil, in some monkeys and in rabbits. The possibility of redistribution of cholesterol from blood to organs must be considered, though this must not be extrapolated to humans.

There seems no doubt that PUFAs act as cholagogues, at least in some subjects, and that this contributes to the reduction of plasma cholesterol. The enhanced catabolism of cholesterol by PUFA and the inhibition of reabsorption of cholesterol from the intestinal lumen by the sitosterol in the oils may ultimately prove to be the two decisive phenomena responsible for the hypocholesterolemic effect of vegetable oils. The question is open.

E. Phytosterols-Sitosterol as Cholesterol Depressants

'Plant sterols', 'soybean sterols' and 'corn oil sterols' have often been used·in animal experiments and in clinical studies without disclosure of the total sterol content or sterol partition of such preparations. Where the composition has been revealed, the β-sitosterol content turned out to be often higher than in some preparations which bore the label 'sitosterol'. Thus, all reports on 'phytosterols' and on 'sitosterol' are considered under a single heading.

Vegetable oils were used first for human trials, and only later in animal experiments. With regard to the study of phyto- and sitosterol the sequence was reversed. *Peterson* [646] reported in 1951 on the effect of 'mixed soy sterols' on the plasma cholesterol of chickens. *Pollak* [660] conducted experiments in 1950–1951, feeding rabbits with 'sitosterol'. Results of these experiments formed the basis for sitosterol administration to humans. His first report, in 1952 [660], referred to both species. For historical reasons, we will first deal with chickens and rabbits, then with other animal species, in alphabetical order, and later, with essays on human studies. Consideration will be given to the factors which influence the outcome of trials, to be absorbability of sterols, and to the concepts concerning the mode of action of sitosterol.

1. Animals

a) Chickens

In all studies with chickens, the birds were either rendered hypercholesterolemic prior to feeding of plant sterols, or sitosterol and cholesterol were fed together in various proportions.

Peterson [646, 647] and *Peterson et al.* [648–652] fed chickens a diet supplemented with 0.5–1% cholesterol, 0.5–1% mixed soy sterols in 4–4.5% cottonseed oil, or both sterols. After 10 weeks on a basal ration, the plasma cholesterol averaged 196 mg/dl, the liver cholesterol, 9.69 g/100 g dry weight. Addition of 1% cholesterol to the basic ration changed the two levels to 942 mg/dl and 9.46 g/100 g. On further addition of 1.3 g soy sterols to the diet, increase in plasma cholesterol and deposition of liver cholesterol were inhibited. In another study, the two sterols, when fed separately, led to a 133-mg/dl plasma cholesterol level, when fed together, to a level of 109 mg/dl. Steady plasma and liver cholesterol levels of hypercholesterolemic chickens could be maintained by adding 1.3% soy sterols to the diet [646]. When chicks were fed a diet containing either 1% cholesterol or 1% of several cholesteryl esters, only cholesteryl acetate caused a degree of hypercholesterolemia comparable to that induced by cholesterol. Liver cholesterol was also markedly increased. The cholesterolemic effect of other esters was: oleate > caprate > myristate > palmitate > stearate.

These results served as a comparison with the effect of β-sitosterol, stigmasterol and ergosterol, all of which prevented a rise in plasma cholesterol. On a basal diet of 0.05% cholesterol, 0.02% of any plant sterol, and 43% crude fat for 2–3 weeks, the effect of the phytosterol supplement was minimal. However, when 0.25–3% soy sterol was fed, both plasma and liver cholesterol decreased. Soy lecithin was not effective [646]. The condensed reports on experiments with chickens are summarized in table XVII.

The chicken is well suited for experiments in which soy sterols or β-sitosterol are used to depress plasma cholesterol. Failure to obtain results can be explained by the very small dose of dietary sitosterol fed [788]. In a variety of experiments, diets containing 1–5% soy sterols fed for 4, 10 or 28 weeks effectively lowered induced hypercholesterolemia from a level of 1,300–1,400 to about 370 mg/dl. Feeding cholesterol plus sitosterol in a 5:1 or 4:1 ratio resulted in plasma cholesterol levels which were lower than levels caused by cholesterol feeding by 11–16%. Where the plasma cholesterol was depressed by 17%, the liver cholesterol was lower by 27%,

Table XVII. Influence of phytosterols on cholesteremia and atherosclerosis in the chicken

Sterol	%	Cholesterol %	Other treatment or prior state	Cholester-emia	Athero-sclerosis	Ref.
Soy	1.3	1	–	inhibited	ND	646
SITO	0.10–0.19	1	–	no effect	ND	788
CSS	5	1	–	inhibited	inhibited	130
Soy	1.3	1	hypercholesteremic	inhibited	inhibited	698
Soy	5	–	hypercholesteremic	reduced	ND	657, 658
Soy	5	–	estrogen	no effect	ND	
SITO	4	2	hypercholesteremic	reduced	reduced	227, 229
Soy	3	1	–	reduced	–	54, 55
Soy	1	1	gallogen	reduced	reduced	172
TOS	1–2	2	–	reduced	ND	721
Soy	1	5	–	reduced	ND	774
Soy	1	1	hypercholesteremic	reduced	reduced	467
SITO	0.5–1		atherosclerosis	inhibited	inhibited	102
COS	0.05	egg powder	–	ND	inhibited	277

SITO = Sitosterol; CSS = cottonseed oil sterols; TOS = tall oil sterols; COS = corn oil sterols; ND = no data.

the aortic cholesterol by 29%. Since the chicken responds to hypercholesterolemia with atherogenesis, prevention of hypercholesterolemia led to reduction or prevention of vascular alterations – even where such was not reported (table XVII).

b) Rabbits

As in the chicken, altherogenesis in rabbits depends on the level and duration of hypercholesterolemia. Also, as in the experiments with chicken, rabbits were either rendered hypercholesterolemic prior to start of a phytosterol regimen or were fed simultaneously sitosterol and cholesterol. And, as in chicks, prevention of hypercholesterolemia prevented atherogenesis, even when not recorded.

Pollak [662] fed 115 rabbits either 1–1.5 g cholesterol/day (C₁, C₂), 15 ml cottonseed oil in alcohol (B), or 1, 2, 3, 5, 6, 7, or 10 g sitosterol (D₁–D₇) and, in a second study, sito- and cholesterol together in 1:1 (E), 3:1 (F), 5:1 (G), 6:1 (H), or 7:1 (I) ratios for 14 days. The results are seen in figures 4 and 5.

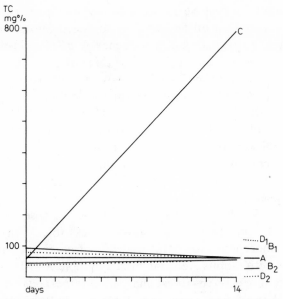

Fig. 4. Total serum cholesterol (mg/dl) in rabbits fed: (A) basal diet; (B) cottonseed oil/95% ethanol 2:1; (C) 1–1.5 g/day of cholesterol in B; (D) 102 g/day sitosterol in B [662].

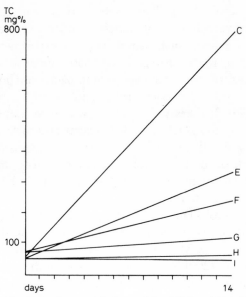

Fig. 5. Total serum cholesterol (mg/dl) in rabbits fed: 1g cholesterol (C); sitosterol: cholesterol (S:C) 1:1 (E); S:C 3:1 (F); S:C 5:1 (G); S:C (H); or S:C 7:1 (I) [662].

Dietary cholesterol given alone caused a 14- to 16-fold increase in serum cholesterol. Cottonseed oil feeding had no effect. Sitosterol, in varying amounts, from 1 to 10 g/day, had no influence. At a S:C ratio of 1:1, the serum level increased sixfold, at S:C = 3:1, threefold, at S:C = 5:1, twofold. At S:C = 6:1, the serum level was 25% above the initial value, and at S:C = 7:1 it remained at the basal level. Cholesterol alone caused extensive atherosclerosis in all rabbits. 2 of 10 rabbits fed S:C at a 1:1 ratio had aortic lesions. In none of the rabbits fed sitosterol in amounts exceeding cholesterol by over 2:1 were there either gross or microscopic lesions. A serum cholesterol lower than 246 mg/dl, such as obtained with a S:C = 3:1 ratio, prevented atherogenesis. Excess of sitosterol over cholesterol needed for results was explained by two factors. One, the 'sitosterol' preparation used contained only 75–80% β-sitosterol; thus, a 7-gram dose corresponded to only 5.4 g of β-sitosterol. Second, sitosterol has to block not only exogenous but also endogenous cholesterol absorption from the intestinal lumen. Experiments with rabbits are summarized in table XVIII.

The reports of failure deserve critical analysis. *Dreisbach et al.* [235] used a S:C ratio of 1:1, insufficient to elicit effect. *Curran* [208] and *Curran and Costello* [209] added either 3 g mixed soy sterols or 3 g dihydrocholesterol three times a week to the food of three groups of rabbits. Their report that atherosclerosis developed in all the rabbits fed soybean sterols but in none of those fed dihydrocholesterol suggests that the two preparations were switched. Not a single investigator who has experimented with phytosterols has noticed arterial lesions, whereas all those who have worked with dihydrocholesterol found it at least as atherogenic as cholesterol. The interconversion of dihydrocholesterol to cholesterol has been proven by *Cook et al.* [196]. It takes place in insects and in humans. Phytosterol absorption is far too small and their excretion too fast to induce vascular alterations. Another criticism of these experiments concerns the dose and the mode of administration of the plant sterols. In the very first reports on chicken [646] and on rabbits [660] it was pointed out that simultaneous intake of sitosterol plus cholesterol causes greater depression of plasma cholesterol than if these sterols are fed separately. Here, 9 g/week of soybean sterols were given to rabbits on alternating days, three times a week in 3-gram doses. Moreover, *Shipley et al.* [764] pointed out that *Curran and Costello*'s [209] finding of soy sterols in the liver of rabbits was based on erroneous methodology and faulty calculations.

Rabbits are well suited, as are chickens, for experiments with dietary phytosterols and sitosterol. An S:C = 3:1 ratio suffices to prevent athero-

Table XVIII. Influence of phytosterols on cholesteremia and atherosclerosis in rabbits

Sterol	Dose	Cholesterol dose	Prior state	Cholesteremia	Atherosclerosis	Ref.
SITO	1 g	1 g	–	O	O	660, 662
SITO	3 g	1 g	–	R	R	
SITO	7 g	1 g	–	I	I	
PS	0.4 g	–	hypercholesteremia	R	R	878
PS	0.8 g	–	hypercholesteremia	R	R	
PS	3 g	1 g	plus Cholagogue	R	ND	897
PS	2 g	1 g	–	R	ND	889, 890
PS	7 g	1 g	–	I	ND	
Soy	1.25 g	1 g	–	R	–	939
Soy	4 g	1 g	–	I	–	
SITO	2%	1%	hypercholesteremia	R	I	761, 762
SITO	0.24%	–	hypercholesteremia	R	O	59, 62, 64
SITO	2%	3%	–	R	R	
DHC	2%	3%	–	R	R	
SITO	2%	1%	–	R	R	580, 582
SITO	3%	1%	hypercholesteremia	R	R	378
SITO	3%	1%	hypercholesteremia	R	I	373
SITO	4%	1%	–	I	I	602
PS	6%	1%	–	R	R	874
Soy	1%	1%	–	O	O	235
Soy	3%	–	–	ND	lesions	208, 209
DHC	1%	–	–	ND	no lesions	

SITO = Sitosterol; PS = phytosterol; DHC = dihydrocholesterol; O = no effect; R = reduced; I = inhibited; ND = no data.

genesis. An S:C = 6:1 ratio inhibited hypercholesterolemia. This effect is evident after a 2-week ingestion of sitosterol. In rabbits rendered hypercholesterolemic prior to the start of a phyto- or sitosterol regimen a dietary supplement of 3% sitosterol caused regression of hypercholesterolemia to about one half of the plasma level. The lack of information on rabbits' arteries in some essays need not be interpreted as failure to counteract atherogenesis by depression of the serum cholesterol. The few adverse reports can be safely dismissed as being based on ill-conceived experiments with erroneously interpreted results. Regrettably, these reports were uncritically cited by many reviewers.

Table XIX. Influence of phytosterols on cholesteremia and atherosclerosis in dogs

Sitosterol	Cholesterol	Other treatment	Cholester-emia	Athero-sclerosis	References
5%			reduced	no data	951
16 g	8 g	propylthiouracil	reduced	no data	762
40 g	10 g	[131]I	no effect	no data	317
3%	5%		no effect	no data	336
45 g	5 g	[131]I; homograft	reduced	inhibited	430

Table XX. Influence of phytosterols on cholesteremia and atherosclerosis in gerbils, guinea pigs and monkeys

Species	Rations	Cholesteremia	Atherosclerosis	References
Gerbil	S 1% : C 1%	reduced	no data	324
Guinea pig	TOS 2% : C 1%	increased	no data	721
Monkey	S (1 mg/day)	no effect	no data	499

C = Cholesterol; S = sitosterol; TOS = tall oil sterols.

c) Dogs

The first report on the effect of feeding sitosterol to dogs was made in 1929, long before sitosterol was used as a cholesterol depressant [951]. The total number of reports on dogs is small and the dogs were usually made hypothyroid prior to the start of the regimen (table XIX).

The results with dogs were disappointing. The only successful report [430], on feeding S:C at a 9:1 ratio, and the resulting reduction of hyper-cholesterolemia by 50% and prevention of atherogenesis, was criticized in *Nutrition Reviews* [626]. The favorable results were thought to be caused by the dogs' anorexia which led to lower caloric and fat intake.

d) Gerbils, Guinea Pigs, Monkeys

Each of these species, gerbils, guinea pigs, and Cynomolgus monkeys, was used for one study only. No attention was paid to the effect of diets on the state of the arteries (table XX).

e) Mice and Rats

Those who studied mice were mostly interested in the influence of a phytosterol diet on the cholesterol content of the liver. Those who studied

Table XXIa. Influence of phytosterols on liver cholesterol levels in mice

Sterol	Dose	Cholesterol	Cholic acid	Liver cholesterol	References
SITO	0.33 g	–	0.33 g	reduced	783
STIG	0.33 g	–	0.33 g	no effect	783
OST	0.33 g	–	0.33 g	reduced	783
SITO	2.5%	–¹	0.5%	reduced	30
SITO	2.5%	1%	0.5%	reduced	60, 61
SITO	5 mg	5 mg	5 mg	no effect	302, 303
SITO	10 mg	5 mg	5 mg	reduced	302, 303
SITO	40 mg	5 mg	5 mg	reduced	302, 303
SOY	20 mg	in 200 mg fat	–	no effect	724, 725
SOY	30 mg	20 mg	–	reduced	724, 725
SITO	5.40 mg	–	–	reduced	154

SITO = Sitosterol; STIG = stimgasterol; OST = ostreasterol; SOY = soy sterol.
¹ Mice were hypocholesterolemic.

Table XXIb. Influence of phytosterols on serum and liver cholesterol levels in rats

Sterol	Dose	Cholesterol dose	Other treatment	Cholesterol		References
				serum	liver	
COS	20%	–	linoleic acid	ND	I	40
SITO	3%	–	–	R	R	912
SITO	3%	1%	–	R	R	18, 19
SITO	3%	–	–	ND	R	16
COS	0.2%	1%	–	R	R	584
SOY	0.2%	1%	–	R	R	584
SITO	0.2%	1%	methylthiouracil	R	R	580
SOY	5%	3%	–	R	R	156
SOY	2%		–	R	I	835
SOY	5%	1%	thiouracil	R	R	95
SOY	20%		lecithin	R	R	95
SOY	0.5–4%	–	cholic acid (1%)	R	R	385, 386
SITO	3–8.5 g	3 g	–	I	R	733

COS = Corn oil sterols; SITO = sitosterol; SOY = soy sterols;
ND = no data; I = increased; R = reduced.

rats shared this interest but they also reported on the response of plasma cholesterol to such diets (table XXIa,b).

In both species, ingestion of cholic acid resulted in an increase in the cholesterol content of the liver, and β-sitosterol counteracted this effect of cholic acid. The results reflected the ratio of the two dietary additives. In 10 out of 12 studies with rats there was a reduction in serum and in liver cholesterol. One research team [40] reported an increased liver cholesterol content on addition of linoleic acid to the sterol diet. Another group [733] recorded elevation of serum cholesterol, concurrent with its reduction in the liver. The reviewer of *Nutrition Reviews* [624] questioned the results of one study in which rats were fed 1% cholesterol, 0.2% thiouracil, and 5% β-sitosterol, on the grounds that the rats were losing weight, and criticized [620] another report because rats had been kept on a low-fat diet prior to the soy sterol diet and had a low plasma cholesterol to start with.

f) Pigeons

A single report was made on the favorable results obtained with White Carneau pigeons with severe atherosclerosis. *Clarkson and Lofland* [174] were interested in the cholesterol content of aortic lesions and in the degree of atherosclerosis. After five 1-year dietary experiments they found that aortic cholesterol decreased most when a 0.05% vanadium sulfate additive was given, but that this did not affect the aortic lesions. With 10% safflower oil, both the aortic cholesterol and atherosclerosis were markedly reduced. The effect of 6.5% safflower oil plus 0.2% β-sitosterol was not quite as good.

Comments on Animal Experiments. In 1955, during a panel discussion on 'Current concepts in the management of arteriosclerosis', comments were made on the favorable effect of sitosterol on plasma cholesterol in experiments. The panel chairman, *Dock* [232], remarked: 'What happens in the chick and rabbit is not important. Very few of our patients have long ears.' One cannot quarrel with the second part of his remarks. The first part of his comment is, however, wrong. In spite of the obvious differences between chickens, rabbits, and humans, these three species respond very much in the same manner to the ingestion of phytosterols or β-sitosterol. This is not true of many other species. Chickens and rabbits responded well unless the dose of sitosterol was insufficient to produce an effect. The few adverse reports were discussed above, following tabulation of results.

The fact that rabbits respond to sitosterol but not to vegetable oils is puzzling. Corn oil, especially when heated, enhances atherosclerosis in rabbits. This may be due to heated fatty acids of the oil. Rats, too, react better to phytosterols than to vegetable oils. There are species differences with regard to bile acid conjugation and they could account for cholesterol redistribution in rats.

The absence of side effects in animals ingesting sitosterol must be underscored. *Chiu et al.* [168] observed 50 rats fed 0.5–1 g/kg β-sitosterol daily and 12 rabbits fed 1–2 g/kg for 8, 12 and 22 months and did not notice any adverse effects. The same was true for patients who consumed 10–18 g/day for up to 7 years. *Bevans and Mosbach* [103] reported the consistent formation of biliary concretions in rabbits fed 0.25% dihydrocholesterol daily for 2–6 weeks and compared this to the absence of concretions in rabbits fed 1% β-sitosterol for 13 weeks.

2. Classification of Hypercholesterolemias

This chapter is inserted between the analysis of experiments with animals and studies on human populations because there is a direct relationship between cause, i.e. the etiology of hypercholesterolemia, and effect of sitosterol intake. The proposed classification should not be confused with the classification of hyperlipoproteinemia. Since we all ingest some cholesterol there is in all subjects an exogenous cholesterol component which contributes to the pool. However, only in those in whom hypercholesterolemia is clearly due to excessive ingestion does this become the dominant factor.

I. Monogenic, primary, endogenous hypercholesterolemia
   1. Essential, familial hypercholesterolemia ($\pm$ xanthomatosis)
     (a) Homozygous
     (b) Heterozygous

II. Polygenic, secondary hypercholesterolemia
   1. Endogenous, metabolic
     (a) Due to hypothyroidism (primary or secondary, $\pm$ myxedema)
     (b) Due to nephrotic phase of chronic nephritis
     (c) Due to cholestasis (cholangiolitic hepatitis, biliary duct obstruction, biliary cirrhosis)
     (d) As part of dyslipemia of diabetes mellitus
     (e) Concomitant to hyperuricemia ($\pm$ gout)
   2. Exogenous
     (a) Parenterally induced (posttransfusion cholesterol thesaurosis)
     (b) Enterally induced (dietary cholesterol hyperalimentation)

Table XXII. Influence of phytosterols-sitosterol on human serum cholesterol levels (mg/dl)

Number	State of health*	Daily dosage**	Duration (weeks)	Serum cholesterol initial, mg/dl	effect, % (subjects)	Ref.
26	G	S (p) (5–7 g)	2	<240	–13% (8)	663
				>240	–32% (18)	193
1	G	S (s) (15 g)	2¹/₂	225	none	
2	NEPH	S (s) (15 g)	2¹/₂	661	–26%	
7	–	S (s) (6–15 g)			–9% (3)	434, 435
					–28% (4)	
7	G	S (p) (6 g)	26	<300	– 7 to 18%	521, 522
4	HC	S (s) (12–24 g)	4–6	>300	– 6 to 25%	
16	CHD	S (p) (25–40 g)	8–26		– 8 to 47%	
11	CHD	S (p) (10–15 g) (LC)	26–34		– 6 to 25%	
2	G	S (s) (5–6 g)	20–84		– 7 to 20%	97–101
16	G	S (s) (15–18 g)	180		–16%	
12	G	S (s) (6–8 g)	8–56		– 7%	236, 237, 239
12	ATH	S (s) (18–25 g)	8–56		–16%	240
10	ATH	S (s) (20–25 g)	52		–17%	
58	ATH	S (s) (9–18 g)	78	>260	–26%	761, 763
17	ATH	S (s) (9–18 g)	78	<260	none	
25	CHD	S (m) (9 g)	21	<300	–14%	49
6	CHD	S (m) (18 g)	21	>300	–34%	
13	PSOR	S (s) (10–12 g)	3–24	274 (avg)	–33% (3)	687
					none (13)	
7	CHD	S (s) (9 g)	8–32		–15%	534
4	XANTH	S (s) (19–53 g)	8–20		– 6%	692
6	CHD	S (s) (19–53 g)	8–20		–10 to 14%	
13	CHD	S (s) (19–53 g)	4–26		– 9%	796
19		SS (5.7 g)	1–4	214 (avg)	–12%	648
6	G	S (s) (12–18 g) (LF)	26			
9	G	S (s) (12–18 g) (HF)	26	293 (avg)	–17%	263–268
15	MI	S (s) (12–18 g)	26			
118	M	S (m) (15 g)	56	260 (avg)	–14%	262
16	MI	S (s) (20 g)	20	325 (avg)	–17% (14)	531, 532
5	G	S (g) (12 g)	4–8		–13%	712

*G = Good; NEPH = Nephrosis; HC = hypercholesteremia; CHD = coronary heart disease; ATH = atherosclerosis; PSOR = psoriasis; XANTH = xanthomatosis; MI = myocardial infarction; M = mentally ill; UW = underweight; OW = overweight; **S = sitosterol; SS = soy sterols; PHS = phytosterols; TOS = tall oil sterols; LC = low-cholesterol diet; LF = low-fat diet; HI = high-fat diet; (g) = granules; (m) = micronized; (p) = powder; (s) = solution.

Table XXII (continued)

Num-ber	State of health*	Daily dosage**	Duration week	Serum cholesterol initial, mg/dl	effect, % (subjects)	Ref.
25	ATH	S (s) (12 g)	16–80		−22% (15)	199
					−10% (6)	
					+21% (4)	
21	ATH	S (s) (10 g)	8		−10%	907, 908
5	UW	S (s) (7.2 g)	26		−36%	685
			52		−44%	
3	OW	S (s) (7.2 g)	26		−27%	
			52		−36%	
20	G	S (s) (20 g)	4		−30%	230
54	G	S (s) (3 g)	0.33		−15 to 20%	438
15	ATH	S (s) (1.2 g)	1–16		−10 to 20% (7/11)	207
16	AGED	S (g) (10 g)	16		−15%	102
–	–	PHS (+) (10 g)	30		−15 to 20%	779
18	ATH	S (15 g)	4		−25% (9)	731
					− 6% (7)	
					+17% (2)	
7	ATH	S (s) (18–25 g)	18–26		− 4%	437
6	HC	TOS (6 g)	2–3	440 (avg)	−25%	890
9		S (6–15 g)	3	272 (avg)	−15% (7)	234
8	ATH	PHS (p) (3 g)	3–6	306 (avg)	−17%	883
10	ATH	S (s) (18 g)	3–15		−18%	931
58	ATH	S (9 g)	1–7		−15% (32)	440
11	ATH	S (2.5 g)	16–104		−21%	37, 38
6	ATH	S (5 g)	16–104	315 (avg)	−23%	
6	ATH	S (6 g)	16–104		−26%	
8	ATH	S (12 g then 3 g)	16–104	400 (avg)	−42%	
12	ATH	S (s) (8–19 g then 6 g)	3	400 (avg)	−29%	177, 178
54	ATH	S (9 g)	4–6		−21 to 60%	509, 510
30	CHD	S (6–9 g)	2–16		−10 to 40%	887, 888
20	CHD	S (9 g)	9		−34%	

*G = Good; NEPH = Nephrosis; HC = hypercholesteremia; CHD = coronary heart disease; ATH = atherosclerosis; PSOR = psoriasis; XANTH = xanthomatosis; MI = myocardial infarction; M = mentally ill; UW = underweight; OW = overweight; **S = sitosterol; SS = soy sterols; PHS = phytosterols; TOS = tall oil sterols; LC = low-cholesterol diet; LF = low-fat diet; HI = high-fat diet; (g) = granules; (m) = micronized; (p) = powder; (s) = solution.

3. Clinical Trials with Unselected Human Populations

The discussion on the effect of sitosterol should follow the sequence in which we listed the various types of hypercholesterolemia. This cannot be done, however, since the vast majority of essays deal with mixed, unselected populations. These comprise healthy subjects with normal plasma cholesterol levels and others with manifest atherosclerosis and mild to severe hypercholesterolemia. Most investigators ignored the etiology of hypercholesterolemia and this is reflected in the results. Therefore, analysis of essays dealing with various types of hypercholesterolemia is deferred.

Analysis of the bibliography on clinical trials with unselected populations is difficult. The list of missing data is discouraging. The list of modifications of clinical studies and of the variables seems endless. Multiple choices have to be faced: the reports could be arranged chronologically, by health or diseases, by age and sex of the subjects, by their initial plasma cholesterol level, by the type of preparation prescribed, the dose and timing of intake of phytosterols or β-sitosterol, by the length of studies, by the results...

Inasmuch as the first report [663] on the use of sitosterol as cholesterol depressant for humans dealt with most of the variables, it may be considered a 'prototype' study. The essays published during the 15 years following publication of the first report are summarized in table XXII. Essays published more recently will be discussed later. (A summary of the chapter will be made after consideration of many factors [part IV] which influence the results of clinical trials, after comparison of sitosterol ingestion with other hypocholesterolemic measures and agents, and a discussion on the mode of action of sitosterol and its influence on cholesterol metabolism.)

In 1952–1953, *Pollak* [660, 663] reported on the reduction of serum cholesterol of 26 unselected, healthy subjects who ate an unrestricted diet supplemented by 5–7 g/day of a coarse, powdered 'sitosterol' preparation. The daily dose was divided into three portions taken during meals. The fact that this preparation contained only 75–80% β-sitosterol should be kept in mind. No distinction was made as to the type of hypercholesterolemia; some subjects had levels below 250 mg/dl. The initial cholesterol level averaged 256 mg/dl for the group; the range was 126–414 mg/dl. After a 2-week sitosterol regimen, the average was 173.5 mg/dl; the range, 131–264 mg/dl. The correlation between the initial level of hypercholesterolemia and the magnitude of depression of this level was clear-cut, in spite of the small number of subjects (fig. 6).

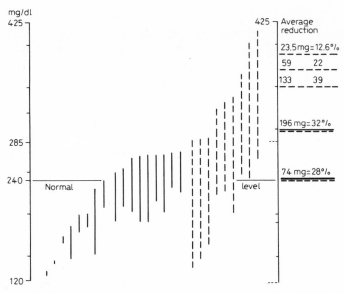

Fig. 6. Effect of sitosterol ingestion on the serum cholesterol level of 26 healthy subjects. Average reduction: 12.6% in normocholesteremia (below 240 mg/dl); 22% in mild hyper-cholesteremia (240–285 mg/dl); 39% in marked hypercholesteremia (285–414 mg/dl); for all levels above 240 mg/dl, 32%; for all subjects, 28% [663].

In those whose serum cholesterol was in the normal range, the effect of the sitosterol regimen was negligible. The effect became statistically significant where it exceeded 10% of the initial value. In all subjects, a second course after a free interval, resulted in about the same effect as the first course. For the whole group, the decrease on the second course was 6% greater than on the first.

One subject, whose serum cholesterol had dropped from 174 to 226 mg/dl on a daily dose of 7 g for 15 days, was after a sitosterol-free period of 40 days, given 5 g/day for 7 days, then, after a 38-day interval, 10 g/day for 64 days. The plotted serum cholesterol values (fig. 7) reveal that the depressed level could be sustained, and they illustrated the speed and degree of decrease in response to treatment and the slope and degree of reversal to the initial level upon discontinuation of the sitosterol regimen.

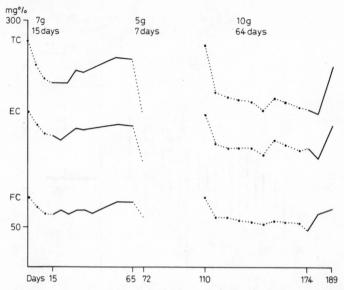

Fig. 7. Response of 1 subject to intermittent (15, 7, and 64 days) ingestion of sitosterol at three levels (7, 15, and 10 g/day). Dotted line = sitosterol regimen; solid line = interval; TC = - total cholesterol; EC = esterified cholesterol; FC = free cholesterol [663].

Curiously, the curves depicted in figure 7 are nearly identical with those shown in an essay by *Mellinkoff et al.* [573], in which they report on the hypocholesterolemic effect in 13 out of 14 patients on a fat-free diet of protein hydrolysate-dextrimaltose. The results of clinical studies with phytosterol-sitosterol are shown in table XXII.

Lack of selection of subjects resulted in the inclusion by several authors of patients with esential hypercholesterolemia or other disorders. These will be reviewed later in the discussion of studies for which only patients with essential hypercholesterolemia, etc., were chosen. Other details, such as the type of preparation used, the duration of studies and maintenance of a lowered cholesterol level will be discussed under special headings. Data from studies in which sitosterol was given in conjunction with other therapy, from studies in which sitosterol was compared to other agents, and from absorption studies could not be included in the table and are discussed elsewhere.

Reports by *Beveridge et al.* [113, 122] are not suitable for tabulation. They fed 85 students a homogenized formula diet with butter as 45% of 980 daily calories for 8 days. Then, 50, 100, 200, 300, 400, 600, 800, 1,600, 3,200, or 6,400 mg β-sitosterol/day were added to the formula diet. The lowest three doses had no effect. At 300 mg or more of β-sitosterol, the plasma cholesterol level decreased by 12.7% ($p < 0.05$–0.02, or 0.01). At 600 mg of sitosterol the decrease was 18.5% and at 6,400 mg it was 35.4%. The reviewer of *Nutrition Reviews* [628] objected to the fact that the whole study lasted only 16 days.

By and large, clinical trials of the first 'sitosterol era' confirmed the original claims, though to different degrees because of modifications of the experiment. The results of European investigators were generally good, possibly because of more frequent use of powdered or granulated sitosterol preparation of higher purity. The report by *Cuny et al.* [207], which heads the list (table XXII) of European authors, is the only one with 'negative' results, an average cholesterol drop by only 3.6%. Most of their patients were pretreated with a low-fat diet and many were given sitosterol for 5, 6, or 10 days only.

Before we analyze the rest of unfavorable reports, it may be well to point out the fallacy of averaging results, especially where a group of patients includes some with monogenic hypercholesterolemia, and diabetics whose plasma cholesterol cannot be influenced substantially or permanently.

Berkowitz et al. [79] arrived at an average 3.7 decrease of the serum cholesterol of 7 men, with the drop ranging from 1 to 24%. *Kaegi and Koller* [437] arrived at an average 18% decrease in 10 men, whose cholesterol dropped from 1.1 to 26.4%.

Billings [128] reported that intake of 12 g sitosterol for 5 months and of 24 g for 2 months had no effect on the serum cholesterol of 1 (one!) patient who during the past 12 years (age, 56–69) had serum cholesterol levels which fluctuated between 238 and 775 mg/dl. The patient, an alcoholic, had mild diabetes mellitus and hypertension and had been treated unsuccessfully with nicotinic acid. Two reports cannot be evaluated for lack of data: *Friedman et al.* [284] treated 6 patients who had coronary heart disease for 6 months with sitosterol, without significant effect on the serum level. No information was given by the authors on the brand, dose, mode of intake of sitosterol, nor on pre- and posttreatment cholesterol levels. *Pomeranze and Chessin* [669] gave to 25 patients, after 4-week control period, 7 g sitosterol (liquified, Merrell) for 6 months without any effect.

The authors also had negative results with a variety of other approaches. Once more, no information was given on the type of patients or on pre- and posttreatment plasma cholesterol levels.

If we dismiss the above-cited reports for lack of information, we are still left with nine 'negative' reports from a single institution. These reports were made by *Wilkinson* alone or with his associates [927–932], or by his associates [140, 141, 537]. All the reports refer initially to 7 patients; 2 more were added later. According to the first report, only 4 of the 7 patients actually completed the prescribed treatment course. Since all the reports arrived at the same conclusions the results can be summarized, in spite of modifications of individual reports. One essay refers to a 5% drop, from 272 to 258 mg/dl, in serum cholesterol of a patient with essential hyper-cholesterolemia and atherosclerosis, on ingestion of 25 g phytosterols from tall oil and soybean oil for 4 months. The selection of patients was not the best; several had essential hypercholesterolemia: of 10 with vascular disease, 9 had angina pectoris; 5 had myocardial infarction, cerebral hemorrhage; 1 or 2 had nephrosis; 2 had xanthomatosis. In some of these patients, the cholesterol level had been reduced before sitosterol was administered. These factors may account for the disappointing results. 1 patient, previously on a low-fat diet, failed to respond to 15 g β-sitoster-ol/day. Another trial, with 15–25 g sitosterol obtained from four different suppliers, failed to lower cholesterol in 5–39 weeks. Two reports stated that after a 10-week diet with vegetable fat, 8 patients received 30–60 ml of a preparation which, according to the supplier contained 16.6% and according to the authors' own analysis, only 14.6–15% soy sterols, princi-pally, γ-sitosterol.

Nutrition Reviews [619] commented on the small doses applied and on the use of γ-sitosterol which, as proved later, contains 7.5–93% cholester-ol. The vegetable oil used by these investigators was 'Ediol', a preparation from coconut oil which has a very low iodine number. The authors themselves were puzzled by their results which were at variance with most other reports. They tried to explain their results by spontaneous fluctua-tions of serum cholesterol which interfere with evaluation of the effect of hypocholesterolemic agents.

These reports were made by a respected clinical investigator and they originated from a prestigious institution. Thus, they were uncritically accepted as authorative and were widely quoted. They became one of several factors which led to discontinuation of sitosterol as cholesterol depressants.

For reasons which will be discussed later clinical studies were resumed in 1974, after a hiatus of nearly 8 years. So far, all the reports, which are more sophisticated than those made years ago, confirmed the cholesterol-lowering ability of β-sitosterol. Because of our mid-1977 cut-off point for the review of the bibliography, several pertinent essays which will appear in print before this book could not be included.

Kudchodkar et al. [511] gave β-sitosterol (Cytellin) to 3 subjects who had a normal plasma cholesterol and to 3 who had hypercholesterolemia. On a 9-gram/day dose, the average decrease was 9% (2.5–13.4%). Fecal excretion of cholesterol was 35–73% higher for those who ingested sitosterol than for controls. *Lees and Lees* [529] stabilized the plasma cholesterol level of 18 patients with familial hypercholesterolemia-hyperlipoproteinemia type II. They used three different sitosterol preparations: (1) Cytellin, a 20% w/v emulsion containing 60–65% β-sitosterol and 40% of other sterols, mainly campesterol; (2) a spray-dried dispersible powder (Lilly) containing 93% β-sitosterol and 7% of other sterols, and (3) the same powder in emulsion form (Lilly). In 9 patients ingesting 18 g of Cytellin daily over a 5-year period, plasma cholesterol decreased by 10.5% and LDL cholesterol, by 15%. 7 patients who responded to 15 g/day Cytellin by a drop of 14% (from 244 mg/dl), reacted to a 5-gram/day dose of the powder with a 16% drop (from 239 mg/dl). In 18 patients, a daily dose of 3 g of the powder resulted in a plasma cholesterol decrease of 9%, and in 8 of these patients a daily dose of 6 g resulted in a decrease of the level by 7%. Emulsified powder was not quite as effective as the powder. Pointing out the low absorbability of sitosterol, the investigators felt that β-sitosterol is a useful adjunct to a diet which is low in SFAs, especially if sitosterol of high purity is used.

Oster et al. [638, 639] treated 40 patients with a diet containing less than 40% fat, less than 300 mg cholesterol and a P:S ratio of 1.5:1. After 8 weeks of placebo treatment the patients received 24 g β-sitosterol in granular form (Lilly) of 94% purity in three equal daily doses taken 30 min before meals for 16 weeks. After elimination of 17 subjects who either had gained over 3 kg or who did not comply with instructions, 23 remained for study. Their plasma cholesterol decreased from 368 to 322 mg/dl, i.e. by 13.2% ($p < 0.001$), and the LDL cholesterol decreased from 231 to 186 mg/dl, i.e. by 19.5% ($p < 0.05$). In 6 subjects, the level fell below the upper limit of the 'normal' range and only in 1 subject did it not fall below the 260 mg/dl mark. Another series comprised 24 subjects who ingested 12 g β-sitosterol daily for 6 months. Their plasma cholesterol decreased by

Table XXIII. Comparison of effect of sitosterol with effects of other agents on serum or aortic cholesterol in experimental animals

Species	Treatment	Serum/cholesterol mg/dl	Ref.
Chicken	1% cholesterol (C)	up to 1,200	467
	1% C + PT (0.5%)	600 (toxic)	
	1% C + CDC (0.5%)	750	
	1% C + SS (1%)	400	
Chicken	1% cholesterol (C)	increased	129
	1% C + MER (0.1%)	rise prevented	
	1% C + CSO	rise prevented	
Pigeon	atherosclerotic	(aortic) increased	173
	VdSO$_4$ (0.05)	no effect	
	DEA (0.05%)	no effect	
	SFO (10%)	reduced	
	SFO (6%) + S (0.2%)	reduced	
Dog	5% cholesterol (C) + ^{131}I	700–930	430
	5% C + ^{131}I + EV (10 mg/week)	500–1,050	
	5% C + ^{131}I + 18:1 + 18:2 (60 g)	630–1,420	
	5% C + ^{131}I + βS (45 g)	< 440	
Rabbit	cholesterol (C) (150 mg/day; 120 days)	430	458
	C + lecithin (2% PhS) (5 g/day)	210	
	C + lecithin (2% PhS) (1 g/day)	300	
	C + choline-HCl (195 mg/day)	340	
Rabbit	cholesterol (C) (500 mg/day; 45 days)	800	897
	C + PhS (s) (3 g/day)	90–270	
	after C-PhS (S) (3 g/day)	358	
	after C-ATT (3 mg/day i.m.)	320	
	after C-no treatment	120–380	
Rabbit	0.5% cholesterol (C) + MTU (5 mg/100 g)	increased	582, 583
	1% C + 2% βS	reduced	
Rabbit	1% cholesterol (C) (14 weeks)	increased	378
	2% C + 2% S	increase inhibited (50%)	
	2% C + 2% S + 2% taurine	no effect	
	2% C + 2% S + 2% glycine	no effect	
Rat	2% cholesterol (C) (14 weeks)	increase	378
	2% C + 4% taurine	increase inhibited (50%)	
	2% C + 4% glycine	no effect	

PT = Phenyltyramine; CDC = α-4-dimethylamine salt of mono (+) camphoric acid; SS = soy sterols; MER = triparanol; CSO = cottonseed oil; DEA = diethanolamine; SFO = safflower oil; EV = estradiol valerate; βS = β-sitosterol; PhS = phytosterol; ATT = anetholtrithione; MTU = methylthiouracil; S = sitosterol; SPL = soy phospholipid.

Table XXIII (continued)

Species	Treatment	Serum/cholesterol mg/dl	Ref.
Baboon	stock diet	103	407
	beef tallow (BT)	391	
	BT + clofibrate (0.3%)	337	
	BT + clofibrate (1%)	156	
	BT + βS (0.5%)	302	
	BT + SPL	263	

PT = Phenyltyramine; CDC = α-4-dimethylamine salt of mono (+) camphoric acid; SS = soy sterols; MER = triparanol; CSO = cottonseed oil; DEA = diethanolamine; SFO = safflower oil; EV = estradiol valerate; βS = β-sitosterol; PhS = phytosterol; ATT = anetholtrithione; MTU = methylthiouracil; S = sitosterol; SPL = soy phospholipid.

25% (p < 0.01), and their LDL cholesterol, by 24% (p < 0.05). The lowered levels were maintained throughout a 1-year observation period.

Success or failure of clinical trials depends largely on the selection of subjects, which obviously should be made in advance, and on the physical and chemical properties of sitosterol preparations.

4. Sitosterol Compared with Other Agents

Parallel studies, in animals and in humans, were made to compare the hypocholesterolemic effect of sitosterol with that of other agents. Pertinent animal studies are summarized in table XXIII.

a) Animals

Discussing the different response of rabbits and rats [378] to bile acid feeding, *Nutrition Reviews* [623] suggested that the fundamental difference in susceptibility to atherosclerosis between these two species may lie in the control of cholesterol excretion. This sounds attractive and simple. Although the relatively resistant rat's bile acids conjugate with taurine more than with glycine, the bile acids of the atherosclerosis-susceptible rabbit conjugate with glycine. The bile acids of the atherosclerosis-prone chicken conjugate with taurine and those of the atherosclerosis-prone man with both taurine and glycine.

Table XXIV. Influence of sitosterol on serum cholesterol levels of human subjects; comparison with other agents

Number	State of health	Treatment	Duration (weeks)	Effect (subjects)	Ref.
52	G	α-tocopherol (10 mg/day)	1	none	104
		α-tocopherol (200 mg/day)	1	none	
		corn oil	1	−23%	
		βS (7 g/day)	1	−21%	
264	aged	12% 18:2 + βS (160 mg)	3	decrease	219
		38% 18:2 + βS (29 mg)	3	decrease	
–	G	Low fat diet		< 10%	426, 428
49	G	βS (18 g/day)	120	> 10%	
50	ATH	brain extract	120	> 10%	
6	HC	PAE (1.2–2.4 g/day)	20	none	235
9	HC	βS (5–15 g/day)	3	−16%	
43	–	MVH (1 cap/day)	20	none (15) decrease (20) increase (8)	731
1	ATH	choline bitartrate	20	none	927, 928
1	ATH	inositol	20	none	
1	ATH	methionine	20	none	
4	ATH	SBO (15–30 g/day)	35	none	
8	MIX	γS (s) (30–60 ml/day)	35	none	
		βS (initially)		decrease	877
		MER (later)		decrease	
10		clofibrate	24	−20% (5)	46
12		DT₄	24	decrease (6)	
9		neomycin SO₄ (12 g/day)	2	decrease	
8	DIAB	βS (6–8 g/day)	2	−12%	
1	XANTH	low-fat diet		626 mg/dl	144
		then, estinyl (0.5 g/day)	5	456 mg/dl	
		then, heparin (40,000 U/day)	5	624 mg/dl	
		then, S (1.6 g/day)	4	468 mg/dl	
6	HC	placebo	12	352 mg/dl	71, 72
		NA (5–6 g/day)	12	293 mg/dl	
		SFO (3 oz/day)	12	345 mg/dl	
		βS (s) (3 oz/day)	12	323 mg/dl	
9	HLP II	NA + CPIB		none	600
1	HLP II	NA + CPIB + DT₄		none	
1	HLP II	βS (11–12 g/day)	64	−39%	
4	HLP II	βS (5.3 g/day)	64	−26%	

G = Good; ATH = atherosclerotic; HC = hypercholesteremic; DIAB = diabetic; XANTH = xanthomatosis; HLP II = hyperlipoproteinemia, type II; βS = β-sitosterol; PAE = 2-phenylbutyramide; MVH = multivitamin hormone capsules (1mg methyl testosterone, 0.75mg ethylestriol,5-triiodothyronine); SBO = soybean oil; γS = γ-sitosterol; MER = triparanol; DT₄ = D-thyroxine; NA = nicotinic acid; SFO = safflower oil; CPIB = clofibrate.

b) Man

As in animal experiments in which the hypocholesterolemic effect of β-sitosterol was compared with the effect of various drugs, analogous studies with humans led to the conclusion that sitosterol was an equal or better cholesterol depressant than other medications (table XXIV).

5. Sitosterol Combined with Other Agents

Since it has been observed that not all subjects respond to a sitosterol regimen and that those who do react do not do so to the same degree, attempts have been made to improve the effect by combining a sitosterol diet with one or more measures thought to have hypocholesterolemic properties.

a) Animals

In rats, lignoceryl alcohol or 2% tetracosanol, waste products from the cellulose industry, did not abolish the serum cholesterol-lowering effect of β-sitosterol when fed in combination with it, according to *Meshcherskaia et al.* [582, 583]. The results were the same as on feeding β-sitosterol alone. *Nath and Harper* [605] conducted numerous experiments with rats, supplementing the basic diet with 0.5% cholic acid, 1% cholesterol, 0.5–1% corn oil, and 0.1, 0.2, 0.5, 1, 5, or 10% sitosterol in all possible combinations. Without sitosterol in the diet the serum cholesterol averaged 442 mg/dl. Corn oil reduced serum and liver cholesterol mildly, sitosterol reduced them markedly. In either case, reductions were greater where bile acid had been administered. The results with sitosterol alone were better than when it was given with corn oil.

Four-week-old cockerels were fed for a 4-week period with a mash containing 4% cottonseed oil and 1% cholesterol by *Clarkson et al.* [172]. Three groups were fed either 1% soybean oil, 0.05% gallogen, or both. Soy sterol, alone or combined with gallogen, lowered serum cholesterol significantly ($p < 0.05$).

Pollak [664] fed 198 rabbits 1 g cholesterol daily for 14 days; their serum cholesterol level increased sevenfold. Then, either 1–7 g/day sitosterol was added, or 0.5–2 g/day linoleic acid (LA, as ethyl linoleate), or a combination of the two. The same results were obtained with S:C = 7:1 as with LA : C = 11 : 1. In combination, one half of each of the two additives sufficed to achieve the same effect as each substance singly. In spite of 1% dietary cholesterol, the rabbits' serum cholesterol level remained normal on ingestion of 3 g sitosterol plus 2–5.3 g linoleic acid.

b) Man

A low-cholesterol diet combined with 9 g β-sitosterol/day was used by *Schön* [730]. Serum cholesterol decreased by up to 40% in 6–18 months – substantially more than on intake of sitosterol alone. *Fahrenbach et al.* [262] gave 15 g sitosterol, 7.5 g corn oil, or both, to their patients, for 13 weeks. All regimens lowered serum cholesterol to the same degree, from 263 to 235 mg/dl, i.e. by 13.5%. *Farquhar et al.* [263–268] kept 15 men on a constant caloric intake of 2,300/day (with 103 g fat) for 28–52 weeks. Replacement of 12 g animal fat and hydrogenated fat by 11 g safflower oil caused a 20% decrease of plasma cholesterol; replacement by 18 g β-sitosterol/day for 7 weeks, also caused a decrease of 20%. By combining safflower oil and β-sitosterol, the decrease was 32%. *Miller et al.* [592] found safflower oil plus β-sitosterol more effective than the oil alone. *Wells and Bronte-Stewart* [913] reported on 1 patient in whom the hypercholesterolemia induced by dietary egg yolk (55 g/day) was inhibited by concurrent feeding of 3 g of β-sitosterol.

Pyridoxine and vitamin E were tested. *Meltzer et al.* [574] treated 28 patients, all with hypercholesterolemia and a history of myocardial infarction, for up to 12 months. Their lowest serum cholesterol value was considered as the baseline for the study. Each patient received four table-spoons (15 ml) daily, a total of 60 ml/day, of a preparation containing 300 mg plant sterols, 1 mg pyridoxine, 9.8 g safflower oil and 20 g vitamin E. The maximum drop was 40 mg. *Engelberg* [255] gave a preparation containing 9.8 g safflower oil, 1 mg pyridoxine, 0.3 g β-sitosterol, and 200 mg mixed tocopherols to 6 patients for 1–2 months. The resulting reduction of serum cholesterol exceeded that obtained on ingestion of safflower oil alone.

Nicotinic acid, effective in some forms of hypercholesterolemia, was also tried in combination with sitosterol. *Berge et al.* [71,72] treated 5 persons, whose hypercholesterolemia had resisted nicotinic acid medication, simultaneously with 4.8 g nicotinic acid and 4.5 g sitosterol. The resulting decrease of the serum level was 21%, twice that obtained with single preparations. 10 patients received daily doses of 5.3 g nicotinic acid or 1 fl. oz. of a 20% suspension of either safflower oil or sitosterol. These three agents reduced plasma cholesterol levels by 24, 6, and 13%, respectively. In 19 patients, 3 fl. oz. of sitosterol plus 4.3 g niacin reduced the plasma level by 21% in 3 months, whereas niacin alone led to a 17% reduction and sitosterol alone to a 12% reduction. *Breslaw* [144] reported a drop in plasma cholesterol levels by 25% on intake of 4.8 g sitosterol/day for 29 days. Addition of 3 g nicotinic acid/day for another 28 days, or of 2 g dessicated thyroid/day for 28 days, had no further hypocholesterolemic effect.

Mugler [599] prescribed 6 g β-sitosterol/day concurrent with mineral waters which, through their $CaCO_4$ and $MgSO_4$ content, act as chola-gogues. In 28 patients, the serum cholesterol decreased by 15–17% within 3 weeks; in 4 patients, by 13% in 4 weeks. In 11 patients whose serum cholesterol was above 330 mg/dl, the drop was by 21%; in 17 others, whose level was below 330 mg/dl, the decrease was only 13%. *Farquhar and Sokolow* [265, 266] treated 11 men, who had atherosclerosis, for 7 weeks with placebo and for 10 weeks with 225 mg gallogen, a choleretic, and 18 g sitosterol/day. 5 other patients received, after the placebo period, 27 g sitosterol only, for an 8-week period. There was a prompt and sustained reduction of serum cholesterol by 16% and of β-lipoprotein by 25% on either diet.

Horlick [402] used β-sitosterol in conjunction with cholestyramine. Both reduced plasma cholesterol by blocking the absorption of cholesterol from the intestinal lumen. Interference with enterohepatic cholesterol flow led to interference with the negative feedback, since it stimulated cholesterol synthesis.

Beveridge et al. [118] fed healthy students corn oil plus β-sitosterol for 8 days. In combination, the decrease of plasma cholesterol was only 1.8% better than on corn oil alone. However, the decrease occurred sooner than when corn oil alone had been taken. *Beveridge* [105] based a patent appli-cation on the postulate that food with at least one edible fat in homoge-neous admixture with 0.75–10% of excess plant sterols, chiefly β-sitosterol, be prescribed. *Campagnoli* [159], reviewing several hypocholesterolemic substances, came to the conclusion that the best therapeutic approach would be a combination of vegetables, cereals, cellulose and sitosterol.

6. Factors Influencing the Results of Clinical Trials

Efforts have been made to explain discrepancies in experimental results. There are many sources of variables. The medication could and should be standardized, i.e. its physical and chemical properties, dose, and mode of administration. To have uniform subjects is not feasible but to select subjects by etiology of hypercholesterolemia is possible and should be the rule. Authors should make available proper data on their patients. To control the investigators' urge for originality may be desirable but not feasible. Those who first reported the hypocholesterolemic effect of soy sterols and of sitosterol had done so after careful studies: they concluded that an excess of phytosterols over dietary cholesterol is required for the effect. Thus, those who used the same preparations originally applied, but who reduced the dosage, should not have been disappointed that their results did not match those of the original claims.

a) Sources of Phytosterols-Sitosterol

Recalling the section on 'Sterols in the Flora', we are reminded of the variation in sitosterol from crop to crop [490], of differences in the relative amounts of sterols in related fruits [604], of uneven distribution of sterols between various parts of the same plant [43], and even between the organelles of plant cells [454]. These are some of the reasons for the variation in products offerred for experimentation with phytosterols or sitosterol. Indeed, the physical and chemical character of β-sitosterol preparations is not uniform, even after purification.

Since β-sitosterol is widely distributed throughout the plant kingdom, various sources have been tapped for commercial-pharmaceutic manufacture. β-Sitosterol from wood shavings proved a better hypocholesterolemic agent than one prepared from soybean [464], which supposedly is 86% pure [799]. Tall oil [317] and cottonseed oil [903] have also been used as sources. Hydrogenation of stigmasterol was said to yield the purest product [83].

Castle et al. [160] found 5.13% campesterol in the commercial 'β-sitosterol' they analyzed. *Auterhoff and Nickoleit* [39] analyzed, by GLC, several commercial preparations of β-sitosterol and found that all had 58–65% β-sitosterol and 37–42% campesterol. *Werbin et al.* [915] analyzed unlabeled samples of 'β-sitosterol' (supplied by Nutritional Biochemical Corp.) and found 4–14% of β-sitostanol. Some of the large number of commercial preparations disclose the composition, others do not (table XXV).

Regrettably, the product used in clinical studies has not always been identified in the reports. Crystalline sitosterol from undisclosed sources was used by at least two researchers [599, 730], and although β-sitosterol was readily available, at least one research team used γ-sitosterol [927].

With the exception of the cholesterol-containing γ-sitosterol, it is difficult to assess the extent to which other sterols in various preparations may have influenced the outcome of studies. The presence of β-sitostanol in a product may not interfere with the results of experiments with rats or may even enhance the hypocholesterolemic effect, according to *Sugano et al.* [816, 817]. Hopefully, similar studies will be made with humans.

Table XXV. Composition of commercial phytosterol and sitosterol preparations

Name	Form	β-Sitosterol %	Other sterols %	
Soybean sterols	p	75	stigmasterol, 10	Distillation Products, Inc.
Soybean sterols	p	54	stigmasterol, 22 campesterol, 24	Distillation Products, Inc.
Corn sterols	–	61	'phytosterols', 7 campesterol, 32	Aldrich Chemical Co.
Anticolesterol	s	20	ND	Pharmacol Inst. Farloni Florence
β-Sitosterol	–	70	campesterol, 30	Sigma Chemical Co.
β-Sitosterol	–	93		Merck & Co.
Sitosterol	g	88		Delande Co., Cologne
Sitosterol	g	40		Boehringer, Munich
Sitosterols	p	70–80		Distillation Products, Inc.
Cytellin	s	60–65		Eli Lilly Co.
Cytellin	s	92	β-sitostanol, 8	Eli Lilly Co.
Positol	p or e	93		Eli Lilly Co.
Sitosol		17	campesterol, 3[1]	Lakemedelsfirmen, Finland

p = powder; s = suspension; g = granules; e = emulsion.

Sitosterol-containing preparations are also available from the companies listed below but they were not tabulated since no data were available: Glidden and Co.; Proctor and Gamble; A.E. Staley Mfg Co.; Aarhus Oliefabrik, Denmark; Promonta, Hamburg; Pharmacol. Inst., Prague; Delande, Paris; Upjohn Co.; Wm. S. Merrell Co.; Schenley Co.; Swift and Co.

[1] Also contains 4% Cognac and 78% water.

Patents. Several patents are registered concerning the preparation and medicinal use of sitosterol.

Yamanouchi [949] obtained a patent for preparation of a mixture of β-sitosterol, campesterol, stigmasterol, dihydrostigmasterol, γ-sitosterol, and dihydro-γ-sitosterol from soybean cakes. *Stern* [800] obtained a patent for mixing equal parts of a soybean preparation which contained β-sitosterol, stigmasterol and campesterol with polyethylene glycol (MW 1,000–6,000). This water-soluble product lowered the serum cholesterol of rats fed rations containing a 1% cholesterol supplement. *Rieckmann and Haack* [691] patented a preparation of sitosterol of small (1–10 μm) particle size for the manufacture of a soluble powder, which could be used in granular form or lyophylized. *Thakkar and Diller* [858] patented the preparation of a dispersible powder of sitosterol. Its high purity (93%) allowed the use of relatively small doses for cholesterol reduction.

Wruble et al. [948] made an edible product by preparing an aqueous emulsion from corn oil. Portions of 15 ml contained, by volume, 33% corn oil, and by weight, 14% β-sitosterol. *Beveridge* [105] based his patent application on his belief that the potency of corn oil as a hypercholesterolemic agent is largely due to its β-sitosterol content. Foods with a hypocholesterolemic potential must contain at least one edible fat with 0.75–10% (by dry weight) plant sterol additive – preferably, β-sitosterol – in excess of that present in the edible fat. *Weigand* [906] obtained a patent based on a combination of β-sitosterol with one of seven bile acids, in a variable ratio. All his drugs could be made into tablets, capsules, elixir, solution, or suspension.

Preparations in which β-sitosterol is combined with other potentially hypocholesterolemic agents seem to offer the best approach to the regulation of plasma cholesterol.

b) Physical Properties

The physical state of sitosterol preparations is an important factor with regard to the cholesterol-lowering effect. Most investigators used suspension containing varying amounts of β-sitosterol. The reason for this lies in the fact that the powdered product originally available was tasteless, chalky and sticky, and not very appetizing. *Pollak* [660, 663] placed the material between two slices of soft bread and ate the sandwich with his meals. Since then, more palatable preparations have become available.

Powdered sitosterol was used by others [434, 435, 883], sometimes in capsules [614], baked into biscuits [49, 533], compressed into tablets [780], or pellets [92, 93] (for rats), or micronized [883] and incorporated into a 'candy bar' [262]. Crystalline sitosterol was fed to mice [952] and rats [584] and to human [599, 730, 733, 734]. Sitosterol granules have also been used [600, 730]. By and large, all these preparations had greater hypocholesterolemic effect than suspensions.

Three teams of investigators compared two or more preparations: granules in tablet form were superior to suspension [102] and a powder also proved more efficient than a suspension although the latter had a higher β-sitosterol content [434, 435]. Lastly, spray-dried powder – even when emulsified – was found more hypocholesterolemic than a sitosterol suspension [529].

c) Timing

For best effect, sitosterol should be ingested simultaneously with meals containing cholesterol, in order to block the absorption of exogenous cholesterol from the intestinal lumen. This was first suggested by *Pollak* [660, 663] and those who adhered to it obtained better results than those who did not. *Beveridge* [104] ascribed his good results with normocholesterolemic subjects to the fact that they ingested β-sitosterol with meals, rather than before meals. The good results reported by *Oster et al.* [638, 639], obtained with small doses of powdered β-sitosterol (93% purity) given 30 min before meals, could quite probably have been further improved by mixing the sitosterol preparation into the food.

The importance of incorporating sitosterol into the food has also been demonstrated in animal experiments. *Peterson* [646] found that soy sterols mixed with the cholesterol fed to chickens was far more effective than when the two sterols were fed separately. *Bartov et al.* [54], also, found that in chickens $S:C = 3:1$ were more effective when fed in mixture than when given separately on alternating days. In rats, *Best and Duncan* [91, 92] did not find any difference in serum cholesterol levels on feeding β-sitosterol-cholesterol together or separately; however, liver cholesterol was depressed more in rats fed the mixture of sterols. *Pollak* [662] fed rabbits daily mixtures of cholesterol and sitosterol – and prevented hypercholesterolemia and atherogenesis; *Curran and Costello* [209] fed rabbits soybean sterols three times a week – and failed.

d) Dosage

The total daily dose of sitosterol is less important than the individual doses taken with meals. Since the amount of cholesterol varies from meal to meal the dose of sitosterol should be proportioned and adjusted with every meal. This is not feasible. The principle of sitosterol 'medication' – if this expression is applicable – lies in blocking the absorption of exogenous, dietary cholesterol. Endogenous cholesterol which is present in the intestinal lumen plays a smaller role, although its amount varies even during 24 h.

Because of differences in demand of the subjects, *Shipley* [763] recommended that the dose of sitosterol be 'titrated' for each individual. This is simple since the effect of ingestion of sitosterol becomes apparent within 2 weeks. Indeed, *Beveridge et al.* [122] have done this and found that the lowest dose of sitosterol which decreased plasma cholesterol to a statistically significant degree ($p < 0.02$), i.e. by 12.72%, equalled 300 mg. *Mühl-*

felder et al. [600] determined the smallest effective dose to be 5.28 g of powdered β-sitosterol. *Audier et al.* [37, 38] switched, after they depressed serum cholesterol levels of their patients by using 12 g/day doses, to a maintenance dose of 3 g/day. *Cloetens et al.* [177, 178] reduced the daily dose of 8–19 g to a maintenance dose of 6 g/day.

e) Duration of Studies – Duration of Effect

One study lasted only 48 h (after a single dose), and one only a single day (with three doses of sitosterol). Other researchers medicated their patients with 2–5 daily doses for periods up to 40 months. Roughly 85% of all clinical studies lasted 8 weeks and 41 lasted 7 months.

Since the effect becomes apparent within 2 weeks, the length of studies is not very important. Practical application and experimental studies have different goals. Sitosterol, like any other dietary or pharmaceutical measure taken to inhibit cholesterol absorption, has to be taken 'forever'. Upon discontinuation of the regimen the plasma cholesterol reverts, as a rule within 2 weeks, to the initial level. Sustaining the depressed level for a period of up to 40 weeks has been reported – once [263]. Substantial (30%) potentiation of the hypocholesterolemic effect of sitosterol on a prolonged regimen has been reported – once. Moderate enhancement of the degree of hypocholesterolemia has been obtained more often, by 'reverse' titration, i.e. increasing the dosage after an initial 2-week trial. The most important question concerns the point at which the hypocholesterolemic effect is nullified by cholesterol biosynthesis. Homeostasis will be triggered by an excess of sitosterol ingestion, by efforts to depress plasma cholesterol below the critical minimal, basal level of an individual. With proper dosage of sitosterol, an excess over the cholesterol pool can be avoided. The equilibrium can be disturbed by interfering illness, also by an excessive cholesterol supply. Here, we may take issue with *Wilkinson et al.* [930, 931] and their 'advice to people, age 45–50, to eat a fat-free breakfast, and anything they want for lunch and dinner'. The impact of even a single overload with fat must not be underrated.

f) Initial Plasma Cholesterol Level

The selection of subjects for clinical study and treatment of hypercholesterolemia is the most important determinant of the results. The correlation between the degree of hypercholesterolemia and the degree of cholesterol depression was pointed out by *Pollak* [663]. As evident from table XXII this observation was confirmed by all investigators who paid any attention to it [37, 38, 49, 98, 152, 193, 434, 435, 440, 509, 510, 599, 692, 761, 880]. *Beveridge* [104, 110] was the only one to obtain satisfactory results in subjects who had a normal plasma cholesterol level.

The inverse ratio between the degree of hypercholesterolemia and the magnitude of the hypocholesterolemic effect does not apply only to β-sitosterol. It is valid for all dietary regimens and many drugs. *Van Handel et al.* [893] found that lecithin is an effective hypocholesterolemic agent, but only for those whose serum cholesterol exceeds 255 mg/dl; certainly not for those whose level is less than 235 mg/dl. *Swell et al.* [840] used similar criteria for the evaluation of the effect of corn oil plus margarine, and *Oshima and Suzuki* [637] used them for safflower oil.

g) Cholesterol Reduction Prior to Sitosterol Intake

The relationship between plasma cholesterol and the effect of sitosterol discussed above explains many of the results from clinical studies. *Tobian and Tuna* [873] depressed the serum cholesterol from 271 to 208 mg/dl by giving a corn oil diet for 4 weeks; subsequent sitosterol intake caused no further drop in the serum level. *Engelberg* [254] made a similar observation in 12 patients whose serum cholesterol had first been lowered by a low-fat diet, when they were given 15–30 g corn oil daily for 3–9 months. The negative results with sitosterol reported by *Wilkinson et al.* [932] could be, at least in part, due to the fact that the serum cholesterol of some patients had been reduced by a low-fat (0.25 g/day) diet prior to sitosterol intake.

The effects of a low caloric, low-fat, low-cholesterol diet, or of a high- or a low-protein diet, etc., either prior to or concurrent with a sitosterol regimen have hardly been studied. The number of pertinent reports is too small to warrant discussion.

7. Age Factor

None of the studies with phytosterols or sitosterol has been designed to deal specifically with the age factor. A few studies refer to weanling, young, and old rats. A few human trials refer to infants, children, adolescents, adults, or the elderly.

Animals. Cholesterol metabolism can be more easily manipulated in young rats than in old. *Wilcox and Galloway* [921] added eight different combinations of 5% lecithin and alfalfa saponins to a ration containing cottonseed oil and 1% cholesterol. On an isocaloric diet, the serum cholesterol of young rats was lower on ingestion of oil than it was on ingestion of lard; in old rats, there was no difference. *Gerson et al.* [296] fed corn oil to rats and noted greater serum cholesterol depression in young rats than in old. According to *Fishler-Mates et al.* [279], cholesterol synthesis in liver slices is reduced with the age of rats (30, 90, and 120 weeks) on soy sterol. *Coleman and Beveridge* [185] found that young rats resisted stress induced by repeated blood letting better than old rats. *Bartov et al.* [55] saw no difference in the response to soy sterols fed to 4-day-old and 3-week-old chickens. *Nesheim et al.* [610], however, reported that soybean meal depressed the fat absorption in two breeds of chicken at the age of 2 weeks but not at 4 weeks.

Man. Man's plasma cholesterol is not static throughout life. The levels fluctuate diurnally, seasonally, and with emotion. Thus, it is necessary to have complete baseline data when evaluating hypocholesterolemic agents. This aspect of cholesterol metabolism will be discussed more fully.

Infants. At birth, the plasma cholesterol level averages 75 mg/dl. It rises in infants fed cow-milk formula during the first months of life, but not in those fed a soy-milk formula, according to *Pomeranze* [668]. He blended corn oil (53 g LA, 12 g SFA, 28 g oleic acid, and 1.5 g/dl sitosterol) into an evaporated milk formula. The serum cholesterol of 53 babies given evaporated milk averaged, after 2, 4, 8, and 12 weeks, 98, 128, 182, and 196 mg/dl, respectively. The corresponding values for 27 babies fed the evaporated milk plus the corn oil formula amounted to 84, 98, 123, and 128 mg/dl, respectively. *Fomon and Bartels* [281] reported that the serum cholesterol level for breast-fed infants averaged 172 mg/dl and for infants fed cow milk 156 mg/dl. Corn oil/coconut oil fed in ratios of 70:78, 57.5:37.5, and 87:13 yielded levels of 143.5, 121, and 108 mg/dl, respectively.

Children. A marked age-correlated response to sitosterol was recorded by *Oster et al.* [639]. 15 adults reacted to 24 g/day β-sitosterol for 4 months with a plasma decrease of 25% but there was only an insignificant drop in 12 children who received 12 g/day. However, two other research teams reported different results. Whereas, by and large, patients with familial hypercholesterolemia responded poorly to sitosterol, *Levkoff and Knode* [539] found that the serum cholesterol of siblings, aged 10 and 11, decreased by 15–20% on three daily doses of 20 ml β-sitosterol (Cytellin). *Segall et al.* [754] gave to 13 children with familial hypercholesterolemia a diet low in SFAs and rich in corn oil; it led to a 25% decrease in serum cholesterol.

Adolescents. In studies on the effect of vegetarian, lacto-vegetarian and non-vegetarian diets on serum cholesterol levels, *Hardinge et al.* [355] and *Hardinge and Stare* [356] found that adolescents did not respond to vegetarian diets whereas in older people who were strict vegetarians serum cholesterol levels were significantly depressed by the diet.

Aging Adults. Reitan and Shipley [688] reported the results of psychologic testing of men of two age groups and observed lowering of serum cholesterol and an improvement in psychologic ability on a β-sitosterol regimen in the 40- to 65-year-old group but not in 25- to 40-year-old group.

8. Sex Factor

While it is seldom considered in animal experiments or in humans, gender may play a role in cholesterolemia and its control.

Animals. Often, age and sex have been mentioned together. Female rats have a higher serum and liver cholesterol than male rats fed rations supplemented with cholesterol. *Okey et al.* [633] and *Okey and Stone* [634] fed hypercholesterolemic rats safflower oil, cottonseed oil, or corn oil: serum cholesterol increased in both sexes but liver cholesterol decreased only in females. Vegetable oils led to an increase in liver cholesterol though to a lesser degree than lard. In females, this difference was not as great as in males. *Roehm and Mayfield* [699] fed 14 male and 14 female rats corn oil and margarine: female rats had a higher serum cholesterol than males. *Gerson et al.* [299] reported that on a fat-free diet with only 0.1 g linoleic acid fed for 1 week to 11 male and 12 female rats there was no difference between the responses. When 2% beef fat was added to their diet, the serum cholesterol of females was much higher than that of males. *Nicolaysen and Ragard* [613] fed two oils for 90 days to rats: on cod liver oil ingestion, the serum cholesterol of females was 194 mg/dl, that of males, 101 mg/dl; on soybean oil intake, females had 122 mg/dl and males had 110 mg/dl. *Katz et al.* [452] obtained different results: after they had fed 3-week-old rats a synthetic diet, they separated them into groups accord-

ing to weight and sex. Some were fed 10% coconut oil, some 0.5% cholesterol, some 1% soy sterols, and others, various combinations of these
additives. After 12 days, plasma cholesterol levels were not significantly
different in any of the groups, but females had consistently lower levels
than males, especially when the diet contained cholesterol. *Rona et al.*
[703] fed rabbits rations containing 95% linoleic acid and an iodine number of 153. They observed a twofold aggravation of atherosclerosis in bucks
and a threefold aggravation in does.

Man. A slight difference in the response to β-sitosterol given to 4 men
and 3 women was noted by *Lesesne et al.* [534]. Initially, all had the same
serum cholesterol level. After 3–8 months, the average reduction of the
level was 16.5% for men, 13% for women. However, the number of subjects
is surely too small to allow conclusions.

9. Weight Factor

No reference to weight has been made in animal experiments and
references to it in reports on clinical trials are rare. *Reeves* [685] observed
that overweight patients whose serum cholesterol averaged 293 mg/dl
reacted to 0.5 oz. (ca. 7.3 g) doses of sitosterol per day with a decrease of the
level to 140 mg/dl, whereas underweight patients whose initial level was
308 mg/dl experienced a reduction to 269 mg/dl. *Meltzer et al.* [574] found
that best results with β-sitosterol are obtained in subjects who did not gain
weight during treatment. *Kaufmann* [453] advised that obese patients with
hypercholesterolemia experience a fall in serum cholesterol whenever
they lose weight but also that there is a tendency for cholesterol to revert to
the original level despite continued therapy. Sparse as these data are, they
may aid in the explanation of the varied responses of patients to a phyto- or
sitosterol regimen.

10. Genetic Factors

The study of genetic factors which could conceivably influence the
response to hypocholesterolemic agents has not yet advanced very far.

Animals. Four reports on experiments with different breeds are available. *Lohman et al.* [548] used pure and cross-bred swine for dietary
experiments. The diets contained 5 or 15% vegetable protein and 50%
calories as corn oil or as hydrogenated soybean oil. The aortae were split
lengthwise. They were first stained with Sudan IV for comparison of

lesions, and then the aortic lipid content was assayed. It was concluded that one half of the variations in lipid content were due to difference in breed. *Corey et al.* [200] compared several species of monkeys fed 10% safflower oil or coconut oil ± 1% cholesterol. The rise in plasma cholesterol was highest in Cynomolgus, next, in Squirrel, and least in Cebus monkeys. *Opdyke and Walther* [636] fed three breeds of 8-week-old chickens 2% cholesterol for 8 weeks, then added 5% refined cottonseed oil to the rations. The percent incidence and degree of atheroma were, respectively, 67 and 1.08% for Kerr-White Leghorn, 65 and 1.07% for Hy-Line White Leghorn, and 17 and 0.35% for Kerr New Hampshire chickens on the two diets. *Subbiah et al.* [808] found that atherosclerosis-prone White Carneau pigeons had more fecal cholesterol and coprosterol and less fecal stigmasterol and β-sitosterol, than atherosclerosis-resistant Show Racer pigeons. These proportions were reversed with regard to the corresponding 5α-stanols.

Man. There are only three references to genetic factors in man. *Connor* [188] stated that neutral fecal sterol and bile acid excretion is enhanced by dietary PUFA in normal subjects and in persons with hyperlipoproteinemia type V, but not in those with type II. *Oster et al.* [638, 639] observed 15 patients of whom 7 had type II B and 8 had type II A hyperlipoproteinemia. Both types responded alike to β-sitosterol. The report by *Segall et al.* [754] is of particular interest since it sheds light on conflicting results in patients with essential hypercholesterolemia and xanthomatosis who received β-sitosterol. They managed to reduce by 25% the serum cholesterol of 13 children with this condition – the children were heterozygous. Children with the homozygous type of essential hypercholesterolemia did not respond.

11. Cholesterol Fluctuation

In the search for an explanation of the disappointing results with sitosterol medication some authors pointed to spontaneous fluctuations of plasma cholesterol. It is not feasible to review the complete literature on this subject. We quote the reports which refer to the use of vegetable oils and of sitosterol, but include only a few where no medication was given at all.

Kritchevsky and McCandless [499] reported spontaneous fluctuation of serum cholesterol during the preexperimental observation period in 6

Cynomolgus monkeys. The oscillations ranged from –27 mg up to +29 mg (average 6 mg). This made evaluation of the effect of 1 g sitosterol (administered daily by stomach tube) impossible.

Discussing the results of clinical trials with sitosterol, *Levere et al.* [537, 538] reported in 9 patients with coronary artery disease serum cholesterol fluctuations of up to +39% and down to –33%, especially in 'older' persons – prior to sitosterol medication. *Wilkinson* [927] observed spontaneous fluctuations in patients with diabetes, constipation, anorexia, or weight loss.

Billings [128] based his criticism of sitosterol as a hypocholesterolemic agent on a 'series of one'. His patient's blood cholesterol ran between 238 and 775 mg/dl over a 12-year period, during which the patient's age advanced from 56 to 68 years. These oscillations were supposedly unrelated to weight and stress. The man was an alcoholic, had mild diabetes and hypertension, and had been medicated unsuccessfully with nicotinic acid. During the year prior to the report, he received 16 g (60 ml) sitosterol daily for 5 months, and then 24 g daily for 7 months. His serum cholesterol values over this period were: 750–233, 500–570, 516, 725–620, and 565–660 mg/dl.

There are individuals who have a stable plasma cholesterol level and others (considerably fewer) who have a labile level. Patients with diabetes mellitus or with recent myocardial infarction are more prone to plasma cholesterol fluctuation than others.

Rivin et al. [694] observed 10 patients, 40–60 years old, who had myocardial infarctions. They were not treated with sitosterol! Duplicate blood samples were sent to five laboratories using, between them, six different methods. There were inter- and intralaboratory differences in the results, with some laboratories reporting consistently higher values than others. The authors concluded that single assays have no value and that a constant decrease by 60 or 100 mg/dl has prognostic value. They failed to point out that none of the six methods used gives consistent results. *Brown and Page* [151] also considered only a cholesterol drop of at least 105 mg to be significant. These demands are unrealistic. They can neither be achieved nor are they desirable unless the degree of hypercholesterolemia is at least 100 mg above the upper normal limit.

Segall and Neufeld [755] stated correctly that the higher the cholesterol level the more likely and the greater are the spontaneous fluctutions. In 1 out of 20 patients the variability reached 61.7% (!) in 2 weeks, 66.5% (!) in 4 weeks, and 95% (!) in 8. In 10 subjects the maximum deviation was

27.7%. Stress usually resulted in a drop but sometimes in a rise of the level. The method used in their assay was not disclosed.

McCann et al. [565] observed variations in patients given two oils (not sitosterol). On peanut oil, serum cholesterol decreased by 31 mg/dl, but then rose by 48 mg/dl, and on sunflower oil it fell by 28 mg/dl, then rose by 43 mg/dl. *Peterson et al.* [654] discussed the cholesterol lability of 'certain individuals' and explained this with regard to individual response to stress and other environmental factors. In 10 patients there were hourly variations. The standard deviations given in their tables seem high: for example 1 subject's cholesterol averaged $263 \pm SD$ 42.8 which would provide a span between a high normal level of 240.5 mg/dl and an abnormal one of 306.1 mg/dl. The method used by these authors has dubious merit. Pessimism concerning these data is based on our experience with multiple methods for cholesterol determination. The digitonin precipitation method gives a complex with many 3β-hydroxysteroids and the colorimetric method is based on the reaction of an acid (almost any strong mineral acid) with a Δ^5-3β-hydroxysteroid.

Keys et al. [460, 461] commented that a decrease of serum cholesterol by 20% exceeds spontaneous fluctuations. Analysis of duplicate samples tested at a 2-week interval (336 analyses in 834 sets from 66 men) revealed in 1,668 pairs a standard deviation of $\pm$ 3.64%. *Farquhar and Sokolow* [264, 265] noted less fluctuation of αLP and βLP-cholesterol during sitosterol administration than during placebo periods.

A report by *Beveridge et al.* [120] concerns data on the serum cholesterol of 93 students on 8- and 16-day dietary regimens. Ingestion of 13–634 mg cholesterol per day caused an appreciable increase in plasma cholesterol. However, ingestion of 1,300 mg cholesterol or more did not further augment hypercholesterolemia. They concluded that each individual has – while healthy – its maximum plasma cholesterol level above which there is no further increase. This matches *Pollak*'s belief that each individual has a minimal basal level below which its plasma cholesterol cannot be depressed without triggering homeostatic reaction. Cholesterol values, given the fluctuations in method and varying conditions of analysis, should not be expressed in fractions of milligrams.

12. Fat Tolerance Tests – Cholesterol Tolerance Tests

In the very first reports it was stressed that β-sitosterol interferes with the absorption of cholesterol. Obviously, this would include LP cholesterol. No suggestion was made that β-sitosterol would in any way affect the metabolism of other lipids. Nevertheless, some reports were made on the reduction of phospholipids and also of total fat, concomitant to reduction of serum cholesterol, as a result of sitosterol ingestion. Nobody suggested reduction of TGs, some even reported a rise in them [96].

Bronte-Stewart and Blackburn [148] reviewed 13 control subjects and 23 patients. The latter had suffered one or more myocardial infarctions 6 months to 2 years prior to the study. 10 of the control group and 15 of the patients were tested repeatedly. Their fat meal consisted of 100 g egg and butter; their medication was corn oil. Total lipids (TL) were assayed for an 8-hour period. Prior to the test meal the peak of the optical density curve in serum was 500 for patients and 450 for controls. The two curves ran parallel. After ingestion of corn oil, the curve of control subjects remained unchanged, that of the patients was markedly flattened, had a much lower peak, and returned much slower to the starting value.

Berkowitz et al. [79–82] published several studies. In 'acute' experiments, 5 subjects with normal serum cholesterol values and 5 patients with hypercholesterolemia and coronary disease were selected for the oral fat tolerance test with ^{131}I-triolein in peanut oil. Ingestion of 6 g (three times 2 g) sitosterol in a 20% suspension had no effect on the elevated and prolonged blood radioactivity of either patients or controls. For 'chronic' studies, they selected 10 patients with a history of cardiovascular disease and serum cholesterol levels between 304 and 455 mg/dl. They received 9 g (three times 3 g) sitosterol daily for 4–6 months. Again, there was no effect on the TG tolerance curve although some patients experienced serum cholesterol depression of 1–42% on a 7-gram daily dose of sitosterol. For this research team the isotope fat tolerance was the only criterion for the diagnosis and grading of atherosclerosis. Naturally, they belittled the value of a sitosterol regimen.

A cholesterol tolerance test was devised by *Kalliomäki et al.* [440] to evaluate the efficacy of sitosterol. The serum cholesterol level of 8 men was assayed at 8.00 and 11.30 hours, once on ingestion of 10 g cholesterol, next on ingestion of 10 g cholesterol plus 6 g sitosterol (S:C = 0.6:1), then on ingestion of 3 g cholesterol, and lastly on ingestion of 3 g cholesterol plus 5 g sitosterol (S:C = 2:1). Cholesterol was consumed in the form of five egg yolks at 08.00, after the first blood sampling. To summarize, on 3 g cholesterol, the average rise of serum cholesterol was 0.4%, on 10 g it was 6.4%. Actually, on ingestion of 3 g cholesterol the level decreased by 1% in 4 subjects, was unchanged in 1, and increased by 12% in 3. On ingestion of 10 g cholesterol, the serum level was unchanged in 4 cases and increased by

12% in 4. The S:C = 0.6:1 formula caused an increase in serum cholesterol by 0.5% in 4 cases, and a decrease by 14.7% in 4. For the 8 subjects, the average drop equalled 3.5%. The S:C = 2:1 ratio caused a rise by 4.3% in 2, resulted in no change in 1 subject, and caused a decrease by 9.6% in 5 subjects. For all 8, the serum level decreased by 9.4% (average). Although this series was too small, a modest trend to a decrease in serum cholesterol level on combined cholesterol-sitosterol ingestion was apparent in one half of the cases.

13. Cholesterol-Sitosterol Analogues

Recognition of the hypocholesterolemic effect of β-sitosterol stimulated interest in the effect of other phytosterols and their esters.

a) Sitosteryl Esters

It has been suggested that sitosterol and cholesterol compete for esterification. One essay, by *Nikuni* [615], anteceded the period. He reported that mice can absorb about 50% of dietary phytosteryl acetate and palmitate. However, parallel experiments revealed that cholesteryl esters were deposited to a great extent whereas phytosteryl esters were stored in small amounts. *Duncan and Best* [237] experimented with sitosteryl esters. Groups of rats were made hypothyroid by ^{131}I and were then fed various diets (table XXVI).

Obviously, only β-sitosterol was effective in counteracting cholesterol influx into the liver (at S:C = 5:1). These results were thought to prove that sitosterol competes with cholesterol for esterification. However, the concept that cholesterol is absorbed only after esterification is not shared by all.

Best and Duncan [88, 89, 91, 92] fed rats for 14 days either cholesterol or cholesteryl esters. The serum cholesterol increased fivefold on a diet containing cholesterol or short-chain esters (acetate and oleate). Feeding of 5% sitosteryl acetate, formate, palmitate, or propionate resulted in 300, 359, 531, and 821 mg/100 g of wet liver, respectively. Liver cholesterol increased equally on ingestion of free sitosterol or its oleate or acetate. The more readily a steryl is hydrolyzed by pancreatic enzyme, the more efficient it will be with regard to reducing liver cholesterol in the rat. Sitosteryl esters were generally ineffective as inhibitors of cholesterol absorption.

Weanling rats were fed 3.2% water-soluble soy-steryl-2-carbamatoglutaric acid, K salts, or the comparable derivative of β-sitosterol, stigmasterol, or campesterol for 4–8 weeks by *Herting and Harris* [384, 385].

Table XXVI. Influence of cholesterol, β-sitosterol and their palmitate esters on liver cholesterol of rats. After *Duncan and Best* [237]

Series	Diets	Liver cholesterol mg/100 g
1	basal (A)	262 ± 27
2	A + 1% cholesterol (B)	1,625 ± 296
3	A + 1% cholesteryl palmitate	357 ± 54
4	B + 5% β-sitosterol (C)	281 ± 28
5	B + 5% β-sitosteryl palmitate	1,259 ± 415
6	C + 3.1% β-sitosteryl palmitate	249 ± 38

The additives caused a reduction of serum and liver cholesterol equal to or exceeding that caused by equivalent amounts of unesterified sterols. Addition of 1% cholesterol to the diet did not alter the results. Soy sterol polyethylene glycol 1,000 succinate was as equally effective as the free sterols, whether or not 1% cholic acid had been added to the rations. These results conflict with the reports that long-chain esters of FAs are ineffective. *Mattson et al.* [561] have recently reported that plant sterol esters effectively inhibit cholesterol absorption in rats. In their studies, β-sitosteryl acetate, succinate, decanoate, and oleate all reduced the absorption of cholesterol from a diet containing a 60–40 mixture of cholesterol and cholesteryl oleate.

b) Sitostanol

Many 'sitosterol preparations' contain a fairly large amount of dihydro-β-sitosterol, i.e. β-sitostanol. A few studies were made with sitostanol alone.

Sugano et al. [816, 817] published two studies and claimed that the hydrogenation of phytosterols from corn oil intensifies the hypocholesterolemic potency. The obvious question concerns the difference between dihydrocholesterol and dihydro-β-sitosterol: the first of the two stanols is highly atherogenic for rabbits and chickens. Male rats were fed hydrogenated phytosterols at a 0.1–1% level, with or without 1% cholesterol, for 10–14 days. The starting material was found, by GLC, to be a mixture of 65% β-sitosterol, 31% campesterol, and 4% of other sterols. The authors describe the hydrogenation process used. Sterols and stanols had hypocholesterolemic effect at the 0.5–1% level; the difference in their effect, at

Table XXVII. Recovery of cholesterol and isocholesterol in the serum, liver and feces of rats. After *Tieri and Tocco* [869]

Diets	Cholesterol			Isocholester-ol: feces mg/day
	serum mg/dl	liver mg/100 g	feces mg/day	
Basal (A)	73.3	232	7.5	0
(A) ÷ 1 g cholesterol (B)	135.0	1,069	86.0	0
(A) ÷ 5 g isocholesterol (C)	63.2	383	26.0	761.3
(A) + (B) + (C)	69.0	832	108.1	744.5

1% ingestion, was in favor of stanols (p < 0.02). Next, young rats were fed 0.5–1% β-sitosterol or β-hydroxysitosterol (HS), for 2–4 weeks. Plasma cholesterol decreased more on HS than on β-sitosterol; liver cholesterol increased equally on HS and β-sitosterol. Sitosterol entered the liver, adrenals, aorta and fat whereas HS did not. On cholesterol-free diets fecal excretion of HS was higher (nearly complete) than that of β-sitosterol. This difference was even more marked on diets supplemented with cholesterol. They found the percent sitostanol to be 4.3 in corn oil, 15.9 in safflower oil, 16.9 in soybean oil, 16.9 in coconut oil, and 25 in rice bran oil. These data obscure the results with sitostanol since coconut oil is atherogenic in rats whereas the other three oils are not.

c) Isocholesterol

Best and Duncan [90, 93] and *Duncan and Best* [238] fed groups of 6 male rats six different diets. To four of the diets they added isocholesterol, a mixture of C_{30} sterols, at 2% of rations. Cholesterol was added as 1% of the rations. Even at this 2:1 ratio the isosterols did have an inhibitory effect. The best results were obtained with dihydrolanosterol, the next best with agnosterol or isolanosterol, and the least good with lanosterol. Only agnosterol was detected in the liver, in small amounts. With rations containing 5% isocholesterol, 68 mg/dl serum and 3.4 mg/g liver cholesterol were detected, compared to 82 mg/dl and 12.2 mg/g in controls. The authors found β-sitosterol a better inhibitor of cholesterol absorption than isocholesterol. Isocholesterol was also used by *Tieri and Tocco* [869] who fed groups of 5 rats four different diets (table XXVII).

In rats, isocholesterol is apparently rapidly absorbed and rapidly excreted. The variance with previously cited studies with isocholesterol is probably due to the use of a 5:1 ratio of ioscholesterol/cholesterol by the second research team while the first team used a 2:1 ratio.

d) Stigmasterol

In 1937, *Sperry and Bergman* [783] found that the effect of stigmasterol on liver cholesterol of mice was inferior to the effect of β-sitosterol. *Altschul* [22, 23] fed rabbits 0.3 g stigmasterol daily for 73–116 days or, to bypass the digestive tract, treated them percutaneously with 0.1 g sitosterol for 81–163 days. No morphologic alterations developed in either series.

In man, *Thiers* [859, 860] and *Thiers et al.* [862] found 15–22 g/day of stigmasterol just as hypocholesterolemic as 10–15 g of β-sitosterol. *Kinsell* [470] and *Kinsell et al.* [483] gave 1–2 g stigmasterol daily for 12 days to a patient and supposedly reduced his serum cholesterol from 150 to 100 mg/dl. *Eneroth et al.* [252, 253] fed stigmasterol to 2 men and recovered in their excretions a variety of phytosterols and steroids. They obtained similar results with 7 men who had hypercholesterolemia. *Schön* [730] prescribed 20 g stigmasteryl acetate to his patients and reported a 20% absorption of the compound – also, severe diarrhea.

e) Fucosterol; Sargasterol

Kaneda [442] fed rats cholesterol and 8% fucosterol in β-configuration and 59% in α-configuration (at C_{20}) which he obtained from brown algae *Fucus gardneri* (β) and from *Sarganum muticum* (α). Rat liver cholesterol decreased, but only up to 14 days of feeding. *Reiner et al.* [686] fed young chicks 1% cholesterol to produce hypercholesterolemia and fatty livers. Concurrent feeding of 1% fucosterol (20-β-methyl-24-methylene cholesterol) brought their plasma cholesterol to a normal level and improved the condition of the liver. Similar experiments with cholesterol plus fucosterol and sargasterol, its α-isomer, were less successful.

f) Lecithin

Although lecithin is chemically unrelated to sterols it must be considered here because lecithin preparations contain phytosterols. *Schettler* [725] reported on the lack of response of 4 subjects, whose cholesterol levels were characterized by great variability, to ingestion of lecithin. *Pottenger and Krohn* [673] worked with 31 control subjects and 91 with hypercholesterolemia. Their total caloric intake was 4,000/day, with 30–40% calories as fat, mainly animal fat. The men received a teaspoon of 'lecithin' – with 29% lecithin, 40% inositol phospholipids, 31% cephalin, and 'a small amount' of phytosterols – for from 4 to 60 months. In 6 men, plasma cholesterol rose by 15 mg, in 13 it fell by less than 15 mg, and in 72 it decreased by more than 15 mg/dl. *Van Handel et al.* [893] gave 16 men for 30–250 days an animal fat diet, then, one of four additives: (a) 2–3 g soya lecithin from 100 g soy beans; (b) 25–50 g peanuts; (c) 25–50 g peas or beans, or (d) whole soybean flour. Only in those whose initial serum cholesterol averaged 300 mg/dl was there a decrease in cholesterol and also in serum phospholipids. In 4 of 6 men who had initially been on a low-fat diet, lecithin proved ineffective.

14. Highly Unsaturated and Low-Cholesterol Eggs

Much effort has been exerted in order to lower plasma cholesterol by non-pharmaceutical means. Vegetarian diets, replacement of animal fat with vegetable oils, food additives (such as egg plant and artichoke), then phytosterols and, prominently, sitosterol were tested as hypocholesterol-emic agents. Attempts have been made to transfer the prevention of hypercholesterolemia from the clinic into the kitchen. The use of vege-table oil margarines with high PUFA content has become popular. Curiously, nobody has yet tried to incorporate micronized β-sitosterol into such margarines and shortenings. Efforts have been made to produce eggs, the major source of dietary cholesterol, which would have a high PUFA content or a low cholesterol content.

The first studies in which eggs were used were made to monitor cho-lesterol transport by its appearance in eggs. These studies will be discussed in the chapter on absorption of sterols.

An attempt to produce eggs with a high PUFA content was made by *Leveille and Fisher* [535]. They fed 12-week-old White Leghorn hens either animal fat (yellow grease) or corn oil or soybean oil plus corn meal, in six different combinations. All the diets contained 50% protein. Six plasma cholesterol levels, corresponding to the six dietary combinations, were 195, 204, 212, 206, 230, and 203 mg/dl. The six iodine numbers of the eggs, again corresponding to the six dietary combinations, were 69, 71, 68, 50, 69, and 80. Plasma cholesterol rose most on animal fat diet and the eggs had the lowest iodine numbers. A lower plasma cholesterol level ante-poned the egg-laying age. There was no significant negative correlation between egg production and plasma cholesterol level. A second attempt was made by *Gordon et al.* [323]. A flock of Black Autrolop hens received a diet containing sunflower seeds and the unsaturation of their eggs increased. *Brown and Page* [152] observed 5 men, aged 26–36 years, whose serum cholesterol averaged 210 mg/dl on an egg-free diet for 18 days. After a week's interval, they ate two eggs daily (PUFA 13%, SFA 35%) for 18 days and the serum level rose to 715 mg/dl. After another week interval, they again ate two eggs each day (PUFA 45%, SFA 35%) for 18 days; then the serum level averaged 845 mg/dl.

Two studies were directed to lowering the cholesterol content of eggs. The first, in which a S:C ratio of 1:1 was used, failed. The second, in which the S:C ratio was increased from 1:1 to 2:1 to 4:1 yielded more favorable results. *Weiss et al.* [910] fed laying hens a diet containing 1% cholesterol. This increased their plasma and egg cholesterol. Addition of 1% sitosterol and 29% safflower oil had no effect. Addition of 1% β-sitosterol and 5% lecithin to a low-fat diet reduced both plasma and egg cholesterol significantly. Combined feeding of cholesterol plus sitosterol increased plasma and egg cholesterol more than when cholesterol alone was fed. Niacin 0.1% had no effect; CPIB (Clofibrate) was toxic and arrested egg production; oral *D*-thyrosine led to reduction of plasma cholesterol but to an increase of egg cholesterol.

Clarenburg et al. [170] fed laying hens a standard diet for 81 days. The egg cholesterol content was constant for each hen but there were significant differences between the birds. The hens received 0, 1, 2, or 4% ^{3}H-sitosterol emulsion in 5% carboxymethylcellulose for 60 days. On ingestion of 1% sitosterol, egg cholesterol decreased by 5.7% after 7–10 days and by 8.2% after 15–30 days. On 2% sitosterol, the decrease was 11% after the first feeding period and 35.5% after the second. On 4% sitosterol ingestion, the respective decreases were 29.5 and 35.8%. Incorporation of sitosterol was 42 mg/egg on 2 or on 4% dietary sitosterol. Egg production was not affected. The authors calculated that a man ingesting 42 mg sitosterol daily (or one of the laboratory-produced eggs) could neutralize 190 mg of his intestinal cholesterol, which is the amount of cholesterol deleted in 21 g of egg yolk.

The laying hen metabolizes dietary sterols in a species-specific and hormone-dependent manner. The hen's metabolism must surely be quite different from that of humans, yet in view of the findings discussed in this section above and their potentially practical implications, further research in this area is certainly warranted.

15. Sitosterol in Various Hypercholesterolemias

Earlier in this text we presented a 'Classification of Hypercholesterol-emias' as a practical basis for the selection of suitable patients for whom a sitosterol regimen offers a chance of success. The subjects who comprised the mixed populations in whom sitosterol had been tried were, to a large extent, those with alimentary hypercholesterolemia. This was reflected in the results, namely in the numbers of subjects who responded favorably.

a) Monogenic, Primary Hypercholesterolemia

This condition is also known as essential, idiopathic familial hyper-cholesterolemia. It is due to an inborn metabolic error. This error is enzymatic and results in excessive cholesterol biosynthesis. Thus, one cannot expect a significant response to hypocholesterolemic measures which are based on the principle of interference with the absorption of cholesterol from the intestinal lumen. Except for the level of hypercho-lesterolemia due to myxedema, the highest plasma cholesterol levels occur in patients with monogenic hypercholesterolemia of the homozygous type. These patients have a high incidence of various forms of xanthoma-tosis and also of atherosclerosis. The latter develops at an earlier age than

Table XXVIII. Sitosterol effect on monogenic hypercholesteremia

Conditions	Change in cholesterol level (patients)	Ref.
Xanthomatosis	no effect	536
Xanthomatosis	no effect	860
Xanthomatosis	significant reduction	434
Xanthoma tuberosum	–25%	144
Xanthoma tuberosum	–20%, transitory (3 of 4)	533
Xanthoma tuberosum	–10 to 14%, transitory	796
Primary hypercholesteremia	no effect	762, 763
Essential hypercholesteremia	no effect	775
Essential hypercholesteremia	–10 to 20%	780
Essential hypercholesteremia and xanthomatosis	–15%	89
Idiopathic hypercholesteremia	–24% (5 of 6)	527
Idiopathic hypercholesteremia	–15 to 20%	539
Idiopathic hypercholesteremia and atherosclerosis	–11 to 22%	534
Idiopathic hypercholesteremia and xanthomatosis	reduction	712

in patients with polygenic hypercholesterolemia. There are exceptions to every rule. The inheritance may be homozygous or heterozygous. The heterozygous form may respond better than the homozygous form. This has been reported only once [754]. Apparently, most investigators who experimented with sitosterol did not distinguish between homo- and heterozygous types. Thus, all the studies are considered together (table XXVIII).

In closing this chapter we recall the reports on patients with familial hypersitosterolemia and xanthomatosis.

b) Polygenic, Secondary Hypercholesterolemia; Endogenous, Metabolic Types

The bibliography on sitosterol medication of polygenic hypercholesterolemia of endogenous, metabolic origin is sparse. The reason is obvious: correction of the underlying diseases will result in the reduction of the plasma cholesterol level. Again, there are exceptions: ingestion of sitosterol together with bile acids seems indicated for patients with biliary cholesterol concretions. Intake of sitosterol, together with cholagogic mineral waters used in the treatment of gout, may be effective.

Hypercholesterolemia in Hypothyroidism. Because of the extremely high plasma cholesterol levels encountered in patients with myxedema the possibility of lowering these levels deserves attention. It is not known whether patients with primary hypothyroidism and those with hypothyroidism secondary to pituitary gland dysfunction respond alike to β-sitosterol. Those in whom a hypothyroid state had been induced by ^{131}I respond with a decrease in plasma cholesterol comparable to that of euthyroid patients. All the pertinent reports were made by *Best and Duncan* and co-workers during 1955–1957. Some of their studies dealt with rats, some with humans, and some with both species.

Rats. Rats fed 1% cholesterol experience within 2 weeks a fourfold increase in liver cholesterol [88–91, 95, 98]. Six groups of rats were kept on a low-iodine intake for 3 days and were then made hypothyroid by i.p. injection of 875 μ Ci ^{131}I per rat. On cholesterol ingestion, the liver cholesterol increased more in the hypothyroid rats than in euthyroid rats. Sitosterol lessened the effect of cholesterol. A semi-synthetic diet which contained 40% butter and 5% cholesterol increased the serum cholesterol of 24 rats to 279 mg/dl, and their liver cholesterol to 128 mg/g within 5 weeks. Microscopically, small lipid aggregates were seen in the walls of myocardial and renal arteries. Addition of 0.2% thiouracil to the diet caused a serum cholesterol level of 1,589 mg/dl and a liver cholesterol of 55 mg/g. One half of these rats had renal infarcts and two thirds had lipoid arterial lesions. On addition of 5% sitosterol to the diet, the serum cholesterol fell to 52 mg/dl and the liver cholesterol to 34 mg/g. No anatomical lesions were found in these rats. The reports moved the referee of *Nutrition Reviews* [624] to inquire about the weight increase of rats fed the various diets and about the possible effect of β-sitosterol on thyroidectomized rats fed an atherogenic diet.

Man. Best and Duncan [87, 94] and *Best et al.* [98, 100] reported on 6 patients, 2 of whom had spontaneous myxedema, 2 had hypothyroidism induced by ^{131}I administration, and 2 were euthyroid. All received 20–25 g β-sitosterol daily for 3–42 weeks. In all patients, the initial mean serum cholesterol level was 367 mg/dl and the final level was 275 mg/dl. 4 patients were rendered hypothyroid by ^{131}I because of congestive heart failure due to rheumatic mitral stenosis. Prior to treatment, their serum cholesterol averaged 146 mg/dl, on administration of ^{131}I, it rose to 196 mg/dl. The final serum cholesterol values in these 4 patients were 136, 147, 149, and 259 mg/dl. A β-sitosterol regimen led to a 26% drop in the fourth patient, the one who had reacted with an increased serum cholesterol. Another patient, with myxedema and myocardial infarction, experienced a marked drop in his serum cholesterol on administration of 18–25 g β-sitosterol/day. With the exception of the above-cited instance, hypothyroid patients responded with an average reduction by 15–16%, i.e. roughly the same as patients with other types of polygenic hypercholesterolemia [100].

Hypercholesterolemia in Nephrosis. The nephrotic phase of chronic nephritis is regularly accompanied by hypercholesterolemia. At necropsy, adolescents are found to have a much higher number of fatty streaks and lipoid lesions than those of comparable age but without renal disease. *Conrad and Furman* [193] reported on 2 patients with serum cholesterol levels of 645 and 758 mg/dl, respectively. On administration of 20 g sitosterol/day for 10 days, these levels decreased by 26% ($p < 0.001$). In 7 healthy persons, with levels between 200 and 250 mg/dl, the decrease was 9%. *Best et al.* [100] reported a drop of serum cholesterol by 15–16% in patients with nephrosis who were on a sitosterol regimen.

Hypercholesterolemia in Cholestasis. Treatment of cholangiolitic hepatitis, and of intra- or extra-hepatic biliary obstruction will, if successful, result in a decrease of the elevated plasma cholesterol level. *Conrad and Furman* [193] reported equivocal changes in the serum cholesterol of 1 patient who had obstructive jaundice, hypercholesterolemia and primary biliary cirrhosis, on ingestion of 20 g corn-sitosterol. *Fink* [274] treated a 73-year-old woman who had an external bile fistula with three daily doses of 7.5 g β-sitosterol (in 20% suspension) for 28 days. There was no definite effect on serum or on bile lipid. Biliary cholesterol fell initially to 6 mg/dl, then rose to 25 mg/dl in 28 days.

Weigand [906] supported his application for a patent for preparations based on the combination of various bile acids and β-sitosterol by presenting his results on 1 patient. A 43-year-old woman ingested 250 mg 3α,7α-dihydroxy-5β-cholanoic acid (CDCA) daily plus 700 mg β-sitosterol thrice a day for 3 months. During this period the size of her gallstones had decreased radiologically by 40%. *Bergmann et al.* [58] found that β-sitosterol enhanced the solubility of cholesterol gallstones. Although the rate of synthesis of primary bile acids, of the extrahepatic acid pool and its components (cholic, chenodeoxycholic and deoxycholic acids) varied from patient to patient, the results on ingestion of 12 g β-sitosterol/day were comparable. In 6 out of 7 patients lithogenicity decreased by 10–15% ($p < 0.005$) in 6 weeks. *Gorolami and Sarles* [295] gave 16 patients 1 g of chenodeoxycholic acid and 13 others 1 g of this bile acid plus 3 g of β-sitosterol. All 29 patients had translucent gallstones and a functioning gallbladder. Treatment lasted from 8 weeks to 40 months (average, 7 months). In the first group, complete dissolution of stones was observed in 2 and partial dissolution in 3 patients. In the second group, complete dissolution occurred in 4 and partial dissolution in 8 ($p < 0.05$).

This phenomenon was also elicited in mice. *Goswanmi and Frey* [325] fed mice eight different diets, including a lithogenic diet of 1.2% cholesterol and 0.5% cholic acid. When 2.5% β-sitosterol was added to this diet the incidence of gallstone formation decreased by 35.5% in male and by 25% in female mice.

Hypercholesterolemia in Diabetes mellitus. Considering the widespread, though quite possibly erroneous belief that diabetes mellitus predisposes to atherosclerosis it is surprising that nobody has made a systemic effort to use sitosterol as a hypocholesterolemic agent in patients with diabetes. Of course, diabetics were included in many of the unselected groups of patients treated with sitosterol, and occasionally this was mentioned in the reports. The only pertinent studies were made by *Kinsell* and his co-workers in the years 1954–1959, at a time when the mode of action of vegetable oils and phytosterols was poorly understood. They used stigmasterol and various sterols and phospholipid extracts of vegetable oils for the treatment of patients with a variety of diseases, including some with diabetes mellitus. The results were not correlated with the clinical diagnosis [469, 471]. However, diabetic patients with advanced vascular disease did not respond to a variety of dietary vegetable material [474]. In 1 patient who had familial hypercholesterolemia and a mild form of diabetes, a controlled diet with a daily ingestion of 101 g vegetable fat and crude soy sterols resulted in a significant reduction of the serum cholesterol. In contrast, purified soy sterol in a 60-g/day dose had no effect. There was prompt rebound to the pretreatment serum cholesterol level on substitution of animal fat for the vegetable fat. At that time, the hypocholesterolemic effect was ascribed to phosphatides in vegetable fats [478]. In elderly patients with diabetes there was no correlation between the degree of atherosclerosis and various plasma lipids [479]. Detailed charts of a 37-year-old diabetic who consumed a non-formula high-vegetable fat diet for over 2 months were presented: serum cholesterol fell from 550 to 150 mg/dl, triglycerides from 710 to 210 mg/dl. This was ascribed to the phytosterol content of vegetable fats [481].

Hypercholesterolemia and Hyperuricemia. Gout and diabetes mellitus, both metabolic diseases, often coexist. Patients with gout and patients with diabetes develop atherosclerosis just as often as individuals who do not have either of these two conditions. A causal relationship between hyperuricemia and atherosclerosis is unlikely. A parallel increase of cholesterol and uric acid has not been proven.

Mugler [599] discussed the dissociation between hyperlipemia and hypercholesterolemia. In this, he found an explanation for the occurrence of atherosclerosis in absence of hypercholesterolemia. He also discussed the hypocholesterolemic effect of $MgSO_4$ and $CaSO_4$ in cholagogic mineral waters ('source hepar'): 3 patients who drank the waters for 3 weeks exhibited decreases in serum cholesterol levels of 15, 16, and 17%, respectively. In 28 patients, some of whom had gout and diabetes, the cholesterol dropped by 22%. This occurred in 85% of the patients. In 4 patients who drank the mineral water and took 6 g/day of β-sitosterol for 7 weeks, the serum cholesterol decreased: from 350 to 331 mg/dl, from 360 to

214 mg/dl, from 320 to 290 mg/dl, and from 310 to 190 mg/dl, respectively. In 11 others, whose cholesterol was above 330 mg/dl (with an average of 356 mg/dl) the decrease averaged 21%, while in 17 patients whose cholesterol was below 330 mg/dl (with an average of 314 mg/dl) the decrease averaged only 11%. The general condition of the patients who had previously complained of vertigo, headaches, general malaise, asthenia, and paresthesia, improved remarkably. In 1 female patient who had been treated with a variety of drugs and diets for a period of 6 years, from the age of 43–49, the combined mineral water-β-sitosterol therapy was the first to cause clinical improvement.

c) Polygenic, Secondary Hypercholesterolemia; Exogenous Types

Under this heading we have placed two entirely dissimilar types of hypercholesterolemia. The first to be discussed is rare, the second frequent. The two types have one thing in common: both are man-made, probably caused be introduction of excessive amounts of cholesterol into the body.

Parenterally Induced Hypercholesterolemia. This type of hypercholesterolemia is discussed for the sake of completeness. Nobody has tried to influence it by a sitosterol regimen and nobody will ever do so. It is not widely known, it is hardly ever recognized during life, and it is never treated. In 1947 *Pollak* [667] observed severe and widespread atherosclerosis at necropsy of a 13-year-old girl. The cause of her demise was generalized purpura due to hypersplenism. In preparation for splenectomy, the child had received 21 transfusions of whole blood during 7 weeks. There was no family history of hypercholesterolemia or of cardiovascular diseases. The plasma cholesterol of the child had not been assayed during life; the postmortem level was 365 mg/dl. Of cource, nobody checked the plasma cholesterol of the 21 blood donors.

Assuming that a blood donor has a plasma cholesterol of 200 mg/dl and that he has a hematocrit of 46% and, further, that the ratio of cholesterol in the erythrocytes to the amount in plasma is 45.6; 100 ml of his whole blood would consist of 54 ml plasma with 108 mg cholesterol and 46 ml of cells with 24.6 mg cholesterol, a total of 132.6 mg/dl in whole blood. 1 U of this blood, i.e. 450 ml, would contain 596.7 or just about 600 mg of cholesterol. Assuming the same hematocrit and the same cell/plasma cholesterol ratio for donors whose plasma cholesterol levels are 240, 250, and 300 mg/dl, respectively, units of whole blood of these donors would provide 720, 750, and 900 mg, respectively. Multiples of 10 of the 4 donors lead to figures of 6, 7.2, 7.5, and 9 g cholesterol. 21 U of 200 mg/dl plasma cholesterol would provide 12.6 g and 21 U of 240 mg/dl would provide 14.7 g cholesterol.

Cholesterol thesaurosis due to multiple transfusions of whole blood is akin to similar iatrogenic hemosiderosis. The postmortem observation stimulated experiments which led to successful production of arterial lesions by the intravascular route in rabbits. Close analysis of atherosclerotic lesions in subjects who had received multiple transfusions confirmed the original observation. Refusal of blood donors with elevated plasma cholesterol levels would eliminate this transfusion hazard. The risk of such transfusion reactions has been minimized by platelet transfusion and erythrocyte transfusion instead of those of whole blood.

Enterally Induced (Alimentary) Hypercholesterolemia. Parenterally induced hypercholesterolemia is preventable by reducing the amount of cholesterol introduced into the circulating blood where it comes directly into contact with the arterial wall. Prevention of alimentary hypercholesterolemia should also be feasible, especially as the amount of dietary cholesterol which is absorbed from the intestinal lumen in a day is not much higher than the quantity supplied by a single blood transfusion. Two approaches are: (1) the reduction of dietary cholesterol (with or without control of total caloric intake, of TGs, FAs, proteins, carbohydrates, and roughage), and (2) ingestion of natural substances or drugs which interfere with the absorption of cholesterol. β-Sitosterol belongs to the agents which reduce or block cholesterol absorption. There are, of course, other approaches to this problem, such as promotion of conversion of cholesterol to bile acids and enhancing the evacuation of the bowels.

Limitation of all these measures lies in the fact that healthy individuals have a physiologic range of plasma cholesterol which can be manipulated only moderately in either direction and, further, in the existence of homeostatic regulation of the cholesterol pool.

Alimentary hypercholesterolemia is the most frequent form of hypercholesterolemias and is present – to various degrees – in 80–85% of all people in the Western World. It is also the type of cholesterolemia best suited for the β-sitosterol regimen. Most of the subjects studied by clinicians belong to this group. Since the reports on clinical trials have been reviewed and tabulated in a preceding chapter there is no need to elaborate further at this point, especially as alimentary hypercholesterolemia will be mentioned again in the summary.

The response of patients with alimentary hypercholesterolemia to sitosterol treatment is statistically significant. The clinical significance has not been established, in spite of the favorable response to sitosterol treatment as discussed in the next section of the book.

16. Effect of Sitosterol on Clinical Atherosclerosis

A chapter on the effect of sitosterol on experimentally induced atherosclerosis is superfluous: it is well known that in three species, chickens, rabbits, and dogs, there is a direct relationship between hypercholesterolemia and atherogenesis. Thus, any mitigation of hypercholesterolemia will also lessen atherogenesis. For man, there is no such linear relation between cholesterol ingestion, plasma cholesterol level, and atherosclerosis.

Concurrent clinical improvement upon plasma cholesterol depression has been reported by some investigators. It strengthens the concept of causal relationships. There are reports on improvement of angina pectoris, of the electro- or the ballistocardiogram, and even on the increase in survival rates.

Several favorable repdorts originated in the Soviet Union. *Vaĭsman and Georgievskaya* [887] and *Vaĭsman et al.* [888] treated 36 patients who had coronary atherosclerosis and angina pectoris with 6–9 g β-sitosterol/day for 2 weeks to 3 months. There was an appreciable decrease in serum cholesterol and a lessening or even disappearance of the angina. It must be added that their patients were also receiving other treatment. *Krivoruchenko* [509] reported on 43 patients with atherosclerosis, some of whom also had hypertension, and on 2 patients with asymptomatic hypercholesterolemia. He later [510] reported on a further 50 patients with coronary artery disease. The daily dose of β-sitosterol was 9 g, the treatment lasted 2–8 weeks. All patients responded with a distinct reduction of the serum cholesterol level, and some also with decreased β-lipoprotein. Their general state of health improved: stenocardia abated, the frequency of attacks decreased, and in 'some cases' attacks ceased.

Encouraging reports appeared in France. *Audier et al.* [37, 38] lowered the serum cholesterol of their patients by 23% on ingestion of 2.5, 5, or 6 g β-sitosterol/day. This, in patients whose pretreatment level averaged 310 mg/dl. In those with 400 mg/dl, the decrease was 44% on a maintenance dose for 4 months and up to 2 years. Menier-type headaches, audiovisual symptoms, dizziness, acroparesthesia, asthenia, and effort angina improved – to varying degrees – in 29 of 60 patients. The improvement was documented by electrocardiography, radiologic examination and blood pressure readings. In 12 patients with hypercholesterolemia and various degrees of clinical atherosclerosis, *Cloetens et al.* [177, 178] recorded an average reduction of 38.5% of serum cholesterol and a subjective clinical improvement. Of 9, the EKG improved in 5, the oscillogram in 6, eye ground changes in 5, and the peripheral circulation in 6. *Warembourg et al.* [905] studied 30 patients, aged from 55 to 80 years. They received 6 g β-sitosterol/day for 21 days, none for 10 days, another 21-day sitosterol course, none for 10 days, and a third sitosterol course for 21 days. Their serum cholesterol decreased and there was clinical improvement. At the onset of observation, all patients had headaches, vertigo, visual changes, walking pains, hypertension, altered EKG and oscillograms. Functionally, 6 improved markedly, 2 moderately, and 5 not at all. The blood pressure fell markedly in 10, moderately in 1, and not at all in 3. In 21 of the above patients tested, the EKG improved only in 1 (the T wave in lead V_6). The oscillogram improved in 21 out of 28, the 'profile' or 'index' improved in 15. 2 patients improved in all respects. *Thiers et al.* [862] added Raynaud's disease and intermittent claudication to the list of conditions susceptible to sito- or stigmasterol treatment. They spoke of a eutrophic vascular action by the phytosterols.

From Scandinavia came a report by *Kaegi and Koller* [437] on clinical improvement in 6 out of 10 patients given 18 g β-sitosterol daily for 3–15 weeks. Angina pectoris improved markedly, but not their EKG. *Turpeinen et al.* [881] compared patients, between the ages of 34–67, in two hospitals for the mentally ill. In one hospital, 354 patients, the control group, received a 'house diet'. In the other hospital, 327 patients received a diet in which milk had been replaced by soybean oil. In the latter group, serum cholesterol decreased, the EKG improved, and coronary mortality was lower. The authors do not mention the phytosterol content of the soybean oil but pointed to the differences in the P:S ratio in the diets.

Meltzer et al. [574] treated 28 patients who had hypercholesterolemia and myocardial infarction. 3 of these patients who had significant angina pectoris improved on a β-sitosterol regimen. *Lehmann* [531] followed 16 patients of whom 9 had myocardial infarction and 6 had angina pectoris, 1 had longstanding but asymptomatic hypercholesterolemia and another had familial xanthoma tuberosum. 2 subjects with normal serum cholesterol levels did not respond to sitosterol; in 14 with hypercholesterolemia the serum level dropped. There was also marked reduction in effort angina of 1 patient, moderate reduction in 2 others. A reversal of abnormally depressed S-T segments in the EKG after Master's exercise test was noted. *Lehmann and Bennett* [532] separated 160 patients who had survived myocardial infarction by more than 30 days into 101 controls and 59 subjects. The latter were given 20 g β-sitosterol daily for 3–36 months. This resulted in significant reduction of serum cholesterol and LDL. There was also a highly significant ($p < 0.001$) improvement in the 3-year survival rate of the treated group.

Kuo [521] reported that in 3 young men with hyperlipemia, all symptoms of atherosclerosis – effort angina and abnormal EKG in Master's test – disappeared on intake of 15–20 g sitosterol/day for 3–6 months. In 2 patients with peripheral vascular disease and intermittent claudication there was a progressive increase in the standard claudication time and in skin temperature readings with the vasodilation test. In 2 patients with coronary and peripheral atherosclerosis and with abnormal ballistocardiograms characterized by very low and bizarre ballistic complexes, there was objective improvement and significant increase in the amplitude of the ballistic waves. All these patients had been given sitosterol for 6–8 months and their serum cholesterol had decreased by 16–28.5%. *Cooper* [199] studied the changes in Gofman's atherogenic index in a group of 25 patients given 25 g sitosterol/day for 4–20 weeks. He found a decrease in the index in 16 patients, no change in 4 and an increase in 5. The serum cholesterol drop averaged 22% in 11 and S_fLP 12–400 decreased by 35% in 14 patients. *Riley and Steiner* [692] and *Steiner and Riley* [796] gave 19–52.5 g sitosterol (Cytellin) to 7 hospitalized patients for 2–5 months. All had coronary atherosclerosis and 3 also had xanthomatosis. Serum cholesterol sterol decreased in all, though only temporarily in the patients with xanthomatosis. In 1 of these, skin lesions became smaller during the treatment. There was no improvement in chest pain or in EKG patterns. According to *Farquhar and Sokolow* [268], combination of 18 g β-sitosterol/day and 81 g safflower oil/day substituted for animal fat, resulted in a 34% reduction of serum cholesterol in 15 patients with clinical evidence of atherosclerosis. No definite change in frequency of angina pectoris was found but they did note in 'some' patients a greater tolerance to exercise.

Certainly, more extensive and protracted studies will be needed before we can arrive at definite conclusions on the clinical response to sitosterol or, for that matter, to any other cholesterol-depressing agent.

As an appendix to the review of reports on the clinical features of atherosclerosis one must mention a report on the effect of sitosterol regimen on mental capacity. It is questionable whether the decline of mental ability with advancing years is in any way related, directly or indirectly, to cholesterol metabolism. *Reitan and Shipley* [688] studied 174 healthy men who volunteered for a 1-year study. The average serum cholesterol level of these men was 240 mg/dl. One third received 0.5 oz. of sitosterol (Cytellin) three times, a day, while two thirds did not. The response varied in regard to serum cholesterol. Psychological testing before and after the 1-year period pointed to a beneficial effect on mental retention and on problem-solving ability in those whose serum cholesterol level had been reduced by more than 10%, regardless of the height of the pretreatment level. This was, however, only true for men aged 40–65, and not for those aged 25–40. The older men scored significantly higher in psychologic tests after 1 year of treatment than prior to sitosterol intake.

17. Conditions Unrelated to Cholesterol Metabolism

It is difficult to speculate on the reasoning (and logic?) behind attempts to reduce the plasma cholesterol of patients afflicted with diseases which are unrelated to disorders of cholesterol metabolism. Nevertheless, such attempts have been made in man and other species.

a) Man

13 20- to 65-year-old patients with psoriasis vulgaris were given 10–12 g of sitosterol emulsion (Cytellin) for a period of 20 days to 6 months by *Reiss and Jaimovich* [687]. Only in 3 patients was there a significant drop in serum cholesterol. 11 had serum levels between 127 and 185 mg/dl, 1 had 214 mg/dl, and 1 had 274 mg/dl. The clinical response was unrelated to the serum cholesterol level; in spite of its reduction from 184 to 114 mg/dl in 1 patient, his skin disease worsened.

In 1955–1961, *Thiers* [859, 860] and *Thiers et al.* [861, 862] ascribed to stigmasterol the property of the 'antistiffness factor', studying this first in guinea pigs, then in man. Thus, they explained the beneficial effect of stigmasterol in scleroderma, cellulitis, phlebitis, certain neuralgias, muscular rheumatism, coxarthrosis, spondylosis, and polyarthritis. Later, they added to this list several vascular and neurovegetative disorders, and also pyodermatitis vegetans. They singled out neuro-epidermoid scleroderma and bone trophism and, later, lupus and neuralgia, as main targets for Thiers' treatment. *Lamberton* [526] treated 50 patients with scleroderma by Thiers' method, using the unsaponifiables of avocado and soybean oils. Within 16–18 months, 80% of patients with scleroderma had benefitted from such treatment, especially those with generalized or disseminated scleroderma. Patients with pigmented or visceral-esophageal scleroderma and those with segmented or localized 'sclerodermiform' states failed to react.

The blood coagulability was not influenced by ingestion of β-sitosterol, according to *Kahn and Munitz* [438] and *Mayer et al.* [562]. It was stated that subjects who ate a diet free of EFAs had marked plasma hypercoagulability.

b) Animals

Experimental Hypertension. Two breeds of rabbit were rendered hypertensive by *Heptinstall and Porter* [373]. Unilateral nephrectomy was followed 2 weeks later by contralateral compression of the renal artery. The rabbits were then fed for 8 weeks with 3.5–4.2 g (average, 4 g) cholesterol/week. Addition of 12 g β-sitosterol (S:C=3:1) led to significant reduction of the hypercholesterolemia and prevention of aortic atheroma and arcus senilis in both breeds. There was no effect on the animals' blood pressure.

Experimental Hyperglycemia. Lin et al. [543] induced diabetes in rats by injecting 50 mg/kg alloxan i.v. at 2- to 3-day intervals. Then, as blood glucose rose to 260 mg/dl, 30 mg/kg alloxan were injected i.v. every other week as a maintenance dose. At this point one set of rats received brassicasterol, another set, β-sitosterol. The sterols were obtained from the roots of sugar cane, used in Taiwan as an antidiabetic drug. No mention was made on the effect on plasma cholesterol. In all rats, both types of plant sterol exerted a hypoglycemic effect which equalled 95% of the insulin strength.

Experimental Gastric Ulcers. Adami et al. [4] induced gastric ulcers in guinea pigs by using histamine. They then tested 42 compounds, administered orally or by s.c. injection for their effect on the ulcers. Some compounds aggravated the ulcers, some were inactive, and some were beneficial. Among the later was β-sitosterol.

Catalase and Peroxidase. Rats and rabbits were fed 1% cholesterol ±2% β-sitosterol (S:C = 2:1) for 40–60 days by *Meshcherskaia et al.* [581].

Cholesterol feeding induced hypercholesterolemia and increased catalase and peroxidase activity in the blood. It also caused liver lipodystrophy, hepatic cellular necrosis, lipoid infiltration of the aorta and the adrenal cortex, and increased resistance to a reduced barometric pressure. Simultaneous ingestion of sitosterol prevented or significantly lessened all the effects induced by cholesterol in rats and rabbits. The blood cholesterol fell from 180 to 90 mg/dl, the catalase activity from 5 to 2.1 mg H_2O_2/0.1 ml.

Thrombocytes. Murphy et al. [602] maintained 4 rabbits on a basic diet ration as controls, and separated 56 other rabbits into groups of 4. Each group received 15 g margarine. In addition to the margarine, 100 mg cholesterol was added to the diet of the second group, 400 mg sitosterol to the third, and 400 mg cholesterol and 100 mg sitosterol to the fourth. In the two groups with sitosterol additive there was a drop in serum cholesterol, but not in the tendency to thrombus formation. Cholesterol, even with the addition of some sitosterol, enhanced thrombus formation in extracorporeal shunts.

Plasma Coagulability. Diller et al. [227] fed 6-week-old cockerels for 2, 4, or 6 weeks on broiler rations which contained 5% cottonseed oil ±1% cholesterol, ±4% sitosterol. Some birds were fed for 2 weeks, others received first cholesterol for 4 weeks, then cholesterol plus sitosterol for 2 weeks. Plasma, liver and aortic cholesterol decreased and an inverse relationship was noted between the degree of lipemia and the plasma clotting time (Stypvén). The coagulation time became normal when the serum lipids were restored to normal.

Phagocytosis. Nicol and Bilbey [612] tested the phagocytic activity of the reticuloendothelial system in 16 rats. Among compounds with estrogenic activity, many steroids depressed the phagocytic index severely (to K = 7 or 8), some only mildly (to K = 13 or 11). In the second group were β-sitosterol and ergosterol; each, on s.c. injection, lowered the index to K = 11.

18. Reviews

In sharp contrast to the many excellent and comprehensive reviews on plant sterols or on insect sterols, the quality of reviews on hypocholesterolemic agents is disappointing. Rarely has a topic challenged so many to write so much and say so little as the topic of regulation of hypercholesterolemia. Practically all the reviews are uncritical compilations of data, and very few are based on the personal experience of their authors.

In all fairness, more than 20 such reviews do not specifically relate to hypocholesterolemic agents or to sitosterol. Rather, they refer broadly to the treatment of atherosclerosis, referring to therapy [283, 661, 784], cholesterol metabolism [641], nutrition [451, 519, 520, 634, 879, 894, 900], dietary fats [14, 24, 397, 546, 570], diets in general [315, 424, 575, 670, 672, 678], and vegetable oils [588]. Not a great deal can be expected from these reviews since references to phytosterols or to sitosterol are made in only passing.

Reviews on the reduction of plasma cholesterol devote anywhere from a single line to half a page to phytosterols or sitosterol. They date from the years 1957 through 1974, and would be expected to become more comprehensive with the growth of the bibliography over these years. This is not so, however. The maximum number of references to sitosterol is found in a review by *Stare et al.* [792] in 1957: they cite 11 'positive' and two 'negative' references with regard to the usefulness of sitosterol. The average number of references comes to only two per review. About one third of all reviews which mention sitosterol cite no pertinent essays at all [1, 3, 6, 24, 25, 73, 84, 141, 159, 169, 198, 201–205, 221, 222, 225, 250, 257, 258, 283, 290, 292, 311, 316, 351, 371, 374, 392, 395, 412, 425, 448–450, 459, 466, 471, 472, 485, 496, 524, 530, 540, 547, 550, 558, 564, 587, 588, 599, 630, 634, 635, 645, 655, 659, 681, 682, 726, 753, 760, 767, 769, 778, 785, 786, 789, 791, 794, 818, 875, 884].

Only a few reviews indicate that the writers had personal experience with sitosterol. *Giacovazzo et al.* [301] referred to their own study of 39 subjects, some with atherosclerosis, in whom ingestion of sitosterol reduced plasma cholesterol from 289 mg/dl to a 200-mg/dl level, and the αLP:βLP ratio from 6.85 to 2.8. *Hermann* [376] and *Hermann and Samawi* [377], reported a 13% decrease in serum cholesterol on 3–6 g β-sitosterol taken 30 min before each of three daily meals by 7 patients for 84 days.

Few reviews mention phytosterols specifically. *Keys and Anderson* [459] made a one-paragraph comment. Those by *Peterson* [647], in 1958, *Portman and Stare* [672], in 1959, *Engelhardt* [256], in 1962, and *Verdonk* [899], in 1961 are satisfactory as far as they go. The last of the above praises sitosterol on the basis of selection of eight favorable reports which contrasted with one adverse report. *Verdonk's* own experience was with 30 patients who responded well to 15–30 g sitosterol/day. He writes about 'a profound favorable change in the biochemical profile of atherosclerosis'. A good review was written in Czech by *Šobra* [779], in 1962, with references to his clinical work and to unpublished experiments with rabbits by *Trčka* [878]. Another good review, in English, came from *Schettler and Sanwald* [727], in 1969.

Brief reviews appeared as editorials [431, 433]. *Juergens and Achor* [436] wrote an editorial review in 1959, with five references. A Russian language one by *Kats* [446], in 1966, cited 18 references. A comprehensive review appeared in 1962 [677], in French, obviously influenced by the enthusiastic reports on sitosterol made by French authors. A less emotional, favorable review was published in German [571], also in 1962, with 30 references. The most recent review on plant sterols – with 43 references – is presented as a book chapter by *Borgström* [138] in 1976.

None of the reviews refers to major adverse effects or side effects of sitosterol, even when taken for a long time. An occasional complaint was made about flatulence. This may have been caused by the vehicle rather than by β-sitosterol. An occasional non-specified and undocumented complaint that 'some patients experienced intolerance' can be – to use legal terminology – dismissed as 'hearsay evidence'.

To say that 'no prophylaxis is possible' and that 'there is no justification even for a special diet' to depress plasma cholesterol since all the known measures 'act in an ephemeral manner and have only psychologic effect' seems entirely out of order. Yet, these are quotes from an essay by *Jimenez-Diáz* [421], written in 1961.

Part III

A. Sterol Absorption and Excretion – Sterol Balance

Clinical interest in β-sitosterol as a hypocholesterolemic agent lasted little over a decade. The reasons for the demise of the first 'sitosterol era' will be discussed later. In the dormant period, before the advent of a new 'sitosterol era', sitosterol was neglected by clinicians. However, it was the subject of many studies by plant physiologists, entomologists, ecologists, and others. Some of this work has been mentioned in the first part of this text.

During the latent period interest continued in the absorption of plant sterols, the effect of these sterols on cholesterol absorption, and in the usefulness of sitosterol as an internal standard in sterol balance studies. The results of this research, particularly those pertaining to man, have a direct bearing on the question concerning the mechanism by which β-sitosterol interferes with the absorption of cholesterol from the intestinal lumen.

The question of absorption of phytosterols fascinated researchers long before the first suggestion was made that these sterols may influence the plasma cholesterol level. Much of the work was done prior to the introduction of isotopes and is thus now obsolete. The second period in this research field can be called the 'isotope era'. The third period, which we call the 'post-isotope era' is characterized by the application of chromatographic methods. In addition to dividing the topic into three parts based on technics, each part is further subdivided into one dealing with animal experiments and another with humans.

1. The Pre-Isotope Era

These studies are largely of historical interest therefore no attempt is made to furnish an exhaustive bibliography.

a) Animals

In 1912, *Ellis and Gardner* [249] reported that phytosterols pass unchanged through the digestive tract of rabbits. *Schönheimer* [736, 737] fed at first 2, later 4, rabbits with sitosterol for 180 and then for 280 days. He found no measurable sitosterol in the carcass. None of his analytic methods would distinguish between various sterols. By fecal analysis, he confirmed [738, 743] the lack of absorption of sito-, stigma-, ergo- and brassicasterol. He and his co-workers [739, 740–742] found no absorption of non-irradiated ergosterol, but marked absorption of irradiated ergosterol. On analysis of the thoracic lymph, they found 0.04% in rats, 0.06% in mice, and 0.055% in rabbits. *Schönheimer and Yuasa* [746] fed 3% sitosterol in 10 ml olive oil/day to 2 rabbits. Their carcasses contained 0.158% cholesterol and those of control rabbits 0.159%. In another experiment, two groups of 9 rats each were fed 0.1 g sitosterol/day for 34 days and 94 days, respectively. The livers contained 0.095% sitosterol (as percent of liver cholesterol). The livers of 9 mice fed 0.03 g sitosterol daily for 30 days contained 0.182% sitosterol, those of 2 cats fed 0.03 g sitosterol daily for 37 days contained 0.485% (as percent of cholesterol).

In 1925, *Van Gierke* [892] concluded from feeding 1 rabbit cholesterol and another rabbit phytosterol (Merck) that both sterols are absorbed but that only cholesterol can be found in the aorta. *Dam and Starup* [211], in 1934, injected i.v. 0.6 g phytosterols (mainly sitosterol) into rabbits for several days. Shortly after the last injection, phytosterol (identified by mp) was present in the liver and the lungs but not in the brain. 1 month later, none was found in the carcass. In dogs fed cholesterol in 5% olive oil portal vein blood cholesterol levels rose by 100, 65, and 47 mg/dl. When 5% sitosterol was added to the diet cholesterol levels decreased by 42, 36, and 33 mg/dl, respectively.

The hen's egg attracted attention as a possible indicator of sterol absorption. *Schönheimer and Dam* [745] found a small but significant amount of ergosterol in the eggs of hens fed this sterol: 0.14%, compared to 0.098% (as percent of cholesterol) in control birds. In 1930–1932, *Page* [640, 641], *Page and Menschick* [642] and *Menschick and Page* [577] recovered 0.5% of ingested non-irradiated ergosterol from eggs from the fourth day of feeding onwards. 1 week after termination of the feeding the eggs had the same sterol content as eggs from control. Years later, in 1966, *Boorman and Fisher* [134] used GLC to monitor β-sitosterol, campesterol and stigmasterol in the feces, tissues and eggs of hens. The hens were fed various diets, including one containing corn oil and one containing soy sterols. On diets containing long- or medium-chain-length triglycerides supplemented with 3% maize sterols for 5 days, the cholesterol content of the eggs was not affected and occasionally traces of β-sitosterol (0.2% of total sterols) were found in the eggs. *Konlande and Fisher* [487, 488] fed hypercholesterolemic chicks soy sterols as 1% of the diet and injected s.c. 20 mg/day of sterols. MS and GLC disclosed similar reduction of liver cholesterol in all the birds. Since the oral and s.c. routes yielded analogous results this was held as evidence of extra-absorptive effects. Intraperitoneal injection of 25 mg/day of soy sterols for 10 days led to reduction of plasma cholesterol by 30–40%, similar to the effect of oral ingestion of soy sterols (with 36% campesterol) or of wheat germ sterols (with 25% campesterol). The results pointed to campesterol as being responsible for the extra-absorptive effect of oral 'sitosterol'.

Another approach to checking the resorption of phytosterols was employed in 1929 by *Schönheimer and Yuasa* [747]. Subcutaneous deposits of sitosterol, or 20/80 stigma-/sitosterol, or of non-irradiated ergosterol remained unabsorbed in a dog. Years later (1964), *Schubert and Rose* [751] produced oil granulomas in mice by s.c. injection of 20 mg Δ^4-cholestene-3,6-dione in sesame oil. Although 1 ml of an ethanol extract of sesame oil contained 5 mg sitosterol none could be extracted from the granuloma; however, 10 mg of cholesterol was found. They thought that, in the absence of any evidence of conversion of β-sitosterol to cholesterol, sitosterol had been absorbed. Dramatic differences in methodology account for the conflicting results of experiments conducted 35 years apart.

Rosenheim and Webster [705] found that rats fed a cholesterol-free diet excreted sitosterol in a manner similar to cholesterol. Levorotatory β-sitosterol fed in Thudichum's phrenosine was excreted as dextrorotatory β-sitostanol which, on oxidation, gave coprositostanone. They also found mp differences between sitosterol from tall oil, and sitosterol from soy bean or wheat germ oil.

Rats also became popular for the study of sterol absorption. Between 1952 and 1955, *Ivy* and co-workers published many papers on this topic, studying rats and also humans [416, 417, 445]. The thesis by *Lin* [544] is representative of all these studies: in rats fed a diet containing cholesterol and oleic acid, serum cholesterol rose from 75–80 to 124 mg/dl. Cholesterol fed with dihydrocholesterol did not increase the serum level (77 mg/dl). Feeding cholesterol plus sitosterol led to a level of 117 mg/dl. However, this combination did not affect the liver cholesterol, whereas cholesterol plus dihydrocholesterol increased liver cholesterol. The conclusion that sitosterol is absorbed by the rat was based on the results of the Lieberman-Burchard reaction. *Schön* [728] stated that 30–60% of 2–4% sitosterol ingested by rats was absorbed, though the liver cholesterol was lower than in cholesterol-fed rats. He admitted that the colorimetric method used did not distinguish between cholesterol and sitosterol.

Roth and Favarger [707] added 13% olive oil or 12% tallow, plus 2.6 g cholesterol, sitosterol, or both sterols to the basic ration of rats for 6 days. In diets containing tallow, 3% [²H]-tripalmitin and 0.75% sterol the digestibility of [²H]-cholesterol was 47%, that of [²H]-sitosterol 28%, and that of [²H]-cholestanol 27%. The cholesterol content of the intestinal wall was, during absorption, thrice as high as the sitosterol content. The coefficient of digestibility of olive oil alone was 94.8–96.1, for olive oil plus cholesterol 94.2–94.4, for olive oil plus sitosterol 95.4–96.9, and for olive oil plus ergosterol 87.9–89.4. However, the digestibility of [²H]-cholesterol plus [²H]-palmitin was 36.3–51.0, of [²H]-sitosterol plus [²H]-palmitin 23.6–30.6, and of [²H]-cholestanol plus [²H]-palmitin 25.7–31.8.

Raicht et al. [679] fed rats a diet containing 0.8% β-sitosterol and 1.2% cholesterol. Cholesterol ingestion enhanced absorption from 1.2 to 70 g/-day. This was compensated for by: (1) inhibition of hepatic cholesterol synthesis; (2) enhanced conversion of cholesterol to bile acids (from 12.7 to 26.3 mg/day), and (3) a slight increase in the excretion of neutral steroids (from 7.7 to 11.2 mg/day). Despite homeostasis, liver cholesterol rose from 2.2 to 9.2 mg/g.

Swell et al. [834] fed rats a purified diet containing 5–25% oleic acid, 1% sodium taurocholate, 2% soybean sterols, or various combinations of these, for 3 weeks. Either oleic acid or bile salts were necessary for the absorption of soy sterols. Liver cholesterol increased on all diets, although the increase was most enhanced in those diets containing soy sterols. Paper chromatography (sensitive to 2.5 µg sterol) revealed no soy sterol in either liver or serum. This was interpreted as being due to very rapid conversion of soy sterol to cholic acid or cholesterol intermediates in the intestine or liver. *Nutrition Reviews* [620] commented that rats had been on a low-cholesterol diet and therefore no further lowering of blood sterols could be expected in these experiments. This criticism also applies to experiments by *Friedman et al.* [284, 285] and *Rosenman et al.* [706], whose rats were starved prior to ingestion of sitosterol. No effect was noted on the absorption of cholesterol from the intestines. Rats were starved before and after thoracic duct cannulation. A diet containing olive oil with cholesterol or β-sitosterol or both sterols was administered by stomach tube. Addition of sitosterol in doses of 15, 25, 50, or 100 mg consistently reduced lymph cholesterol. The reduction was only significant when 25 mg of cholesterol plus 100 mg β-sitosterol (S:C = 4:1) were given. In view of this it is surprising that the authors concluded that 'sitosterol cannot be considered as a truly effective agent in the impedence of absorption of either dietary-derived or intestinally excreted cholesterol in the rat'. In acute experiments, rats were fed a diet containing 100 mg cholesterol plus 100 or 500 mg soybean sterols. Rats fed only olive oil had 75 mg/dl lymph cholesterol between 0 and 24 h and 46 mg/dl between 24 and 48 h. On 100 mg of soy sterol the two corresponding values were 60 and 35 mg/dl and on 500 mg they were 56 and 31 mg/dl. If 75 mg/dl of lymph cholesterol represents 100% absorption, 60 mg would represent 71% absorption (–29%) and 56% would represent 74% absorption (–26%). Then, if 46% represents 100% absorption in the final portion of the lymph, the results after soybean sterol addition would represent decreases of 21% and 32.6%. Their 'chronic' study lasted 3–9 weeks. 7 rats received 2% cholesterol, 1% cholic acid, with or without 2% soybean sterols (as S:C = 1:1 ratio). Serum cholesterol rose to 230 mg/dl. When the diet contained 10% of the soybean sterols (S:C = 5:1), cholesterol was 210 mg/dl.

Gould and Cook [329], reviewing the literature on plant sterol absorption up to 1958, added unpublished information by *Cook and Riddall* on the partial absorption of brassicasterol by the rat.

b) Man

The first report on absorption of plant sterols in humans was made by *Windaus* [935] in 1940. He found phytosterols in the feces of a man who ingested peanut sterols. Many reports on sterol absorption were published by *Ahrens* and co-worders from 1965 to the present time. Although their first report, by *Miettinen et al.* [586], dealt with a single subject, a 39-year-old man with hypercholesterolemia who ingested 40% of 2,400 daily calories in the form of fat for 27 days, the data are of sufficient importance for presentation (table XXIX).

Davignon et al. [213] studied the usefulness of chromic oxide (Cr_2O_3) as an internal marker. They examined 28 patients, in 6 of whom there was a significant inverse correlation between fecal turnover and degradation losses of large amounts of dietary β-sitosterol. It was thought that the results of quantitation of fecal excretion were misleading because the fecal composition presented an index of colonic emptying rather than of colonic contents. The results of balance studies are influenced by enzymatic and

Table XXIX. Intake and output of cholesterol and phytosterols (mg/day) in a man. After *Miettinen et al.* [586]

Δ⁵-Sterols	CH	CA	ST	βS	CS + ST + βS
Intake	71.3	51.9	21.8	220.6	294.3
Output	363.9	47.4	17.3	195.0	259.7
5α-Sterols					
Intake	0.7	5.4	1.0	16.3	22.7
Output	7.1	4.7	1.1	13.7	19.5
Balance	−6.4	+0.7	−0.1	+2.4	+3.2
Total Δ⁵ + 5α + 5β					
Intake	72.0	57.3	22.8	236.9	317.0
Output	371.0	52.1	18.4	208.7	279.2
Net balance	−229.0	+5.1	+4.4	+28.2	+37.8

CH = Cholesterol; CA = campesterol; ST = stigmasterol; βS = β-sitosterol.

bacterial action. *Grundy et al.* [345] reported on the interaction of cholesterol absorption and cholesterol synthesis in man as it influences the miscible cholesterol pool. On ingestion of large amounts of cholesterol, the pool may expand as much as 20 g before the surplus is balanced by fecal neutral sterol excretion. Feedback of cholesterol synthesis is demonstrated on large intake of phytosterols. In the metabolic steady state the absorption mechanism is essentially saturated by large amounts of endogenous cholesterol available for absorption. In 5 men with hypercholesterolemia the absorption of exogenous and endogenous cholesterol was greatly reduced and a compensatory increase in cholesterol synthesis occurred – with 20% representing the limit of effect for both absorption and synthesis.

Abell and Kendall [2] kept patients on a low-fat, cholesterol-free diet which was supplemented daily with 100 g plant sterols. Fecal sterols amounted to 300–900 mg/day. On ingestion of 3.5 g of cholesterol, fecal sterols rose to 700–1,500 mg/day; on ingestion of 0.1 or 1 g of plant sterols, the sterol excretion was 300–900 and 400–1,300 mg/day, respectively. *Connor et al.* [192] studied 6 subjects given a series of diets all of which contained 40% of calories from fat. The first test diet was followed by one high in cholesterol and then by one containing plant sterols at various levels. The three regimes followed one another for a period of 33 weeks. Sterol balance was measured in 7-day fecal pools. On cocoa butter (iodine number 32), serum cholesterol was 222 mg/dl and fecal steroids were

709 mg; on corn oil (iodine number 125), serum cholesterol was 177 mg/dl but fecal steroids were 915 mg. Following the corn oil diet, a second term of cocoa butter diet caused the serum cholesterol to rise to 225 mg/dl and fecal steroids to fall to 629 mg. During these three dietary periods, the shifts in serum cholesterol levels were –1,488 mg while bile acids rose by +1,778 mg from the first to the second period. From the second to the third period serum cholesterol rose by +1,629 mg while bile acids decreased by 2,548 mg. *Connor* [188] pointed out this reciprocity.

Karvinen et al. [445] gave to 16 persons a daily diet containing 10 g animal fat, 10 g vegetable fat, 380 mg cholesterol, and 300 mg phytosterols for 4 weeks. The degree of absorption was similar to the amount synthesized daily. Addition of 1, 3, 6, or 9 g/day of cholesterol to the diet, together with 17–40 g of margarine from hydrogenated soya and cottonseed oil, caused no increment in cholesterol absorption. *Schön* [728–730] studied 2 patients who reacted to sitosterol in different ways. A 33-year-old female received 10 g crystalline sitosterol daily for 7 days. She excreted 72 g of sterols during this period. Her serum cholesterol remained at 172 mg/dl. A 63-year-old man received daily doses of 20 g of sitosterol emulsified in carboxymethylcellulose for 16 days. He excreted 84 g of sterols. His serum cholesterol fell from 237 to 209 mg/dl. The author interpreted these data as evidence of 30% absorption of sitosterol. Admittedly, his calculations were based on the results of colorimetric assays which could not distinguish between cholesterol and sitosterol.

Haust and Beveridge [359] studied 2 men, 1 healthy and 1 a diabetic, who were given a fat-free dietary formula for 8 days, followed by Similac and corn oil which provided 60% of calories equicalorically replacing carbohydrate. Cholesterol excretion increased fourfold to fivefold on the corn oil diet. In the diabetic man, coprostanol increased tenfold. *Nutrition Reviews* [627] criticized these calculations because of the short duration of the study and because fecal cholesterol exceeded seven times the amount of cholesterol left in the plasma pool. Thus, fecal sterol must have originated from tissues or from increased biosynthesis.

2. The Isotope Era

The introduction of isotopes into the research on sterol metabolism, which occurred at about the same time as sitosterol had become popular as a cholesterol depressant, had varied effects. Practically all the animal experiments were performed with rats. Contradictory results led to much confusion and much misinformation about the resorbability of sitosterol,

Table XXX. Recovery (%) of tritium-labeled sterols from erythrocytes and tissues of chickens (recovery after 50 days). After *Kritchevsky and Defendi* [497, 498]

	Sterol		
	[7-^{3}H]-Cholesterol (4.3×10^5)[1]	[G-^{3}H]-Sitosterol (8.2×10^5)[1]	[G-^{3}H]-Lanosterol (8.7×10^5)[1]
Erythrocyte	0.03	0.01	0.05
Brain	1.84	1.56	2.02
Cord	0.50	0.39	0.26
Liver	1.23	2.75	1.83
Spleen	0.40	0.48	0.37

[1] Administered radioactivity (cpm).

especially as some investigators extrapolated the results with rats to humans, and conducted their clinical studies with the goal of confirming their results with rats.

a) Animals

Chicks were used by *Kritchevsky and Defendi* [497, 498] who injected 0.5 ml emulsified 7-[^{3}H]-cholesterol, [7-^{3}H]-sitosterol, or [7-^{3}H]-lanosterol into the egg yolk sac of newly hatched chicks. Radioactivity was demonstrated in the brain within 4 days. Nine weeks after injection, 1 bird in each series was killed and radioactivity was assayed in the non-saponifiables from various sites (table XXX). Radioactivity levels decreased between 85 and 120 days. After 50 days, autoradiography showed that radioactivity in the brain was practically nil. The authors were not able to recover radioactive cholesterol from birds injected with lanosterol.

Chickens were used by two other research teams. *Clarenburg et al.* [170] fed hens a 2% ^{3}H-sitosterol emulsion and found a 60% absorption for laying hens and an 85% absorption for non-laying hens. *Sklan et al.* [774] fed 20-day-old chickens for 7 days on a diet of 0.5% cholesterol ± 1% soy sterols. During the last 4 days, yttrium was added as a non-absorbable marker. Soy sterols reduced the size of the plasma-liver cholesterol pool, decreased the daily net absorption of cholesterol and bile acids in the jejunum, and depressed the daily endogenous secretion into the duodenum. The net cholesterol absorption decreased from the 22% (found on sole cholesterol ingestion) to 6% on a soy sterol-cholesterol diet. The absorption for the four sterols was: 6.6% for cholesterol, 4.7% for campesterol, 1.7% for stigmasterol, and 1% for sitosterol.

For guinea pigs. *Glover et al.* [307] and *Glover and Morton* [308] reported absorption of less than 3% for ergosterol-^{14}C hydrogenated to brassicasterol (7,8-dihydroergosterol). *Schönfeld and Sjövall* [735] monitored cholesterol, coprostanol, cholestanol, β-sitosterol, campesterol, and stigmasterol in the bile of guinea pigs by GLC. The amounts of phytosterols were comparable to those of other sterols, except for a smaller amount of coprostanol.

St. Clair et al. [793] compared three flow markers in three species of monkeys. The markers were given by stomach tube. 4 stumptail monkeys, *Macaca aretoides,* 6 African green monkeys, *Cereopithecus aethiops,* and several Squirrel monkeys, *Saimiri sciureus,* were fed mixtures of sterols. Of the markers, ^{14}C-sitosterol, chromic oxide (Cr_2O_3) and ^{3}H-glycerol triether, the first proved most efficient as it was recovered by 95%. Chromic oxide was recovered at 87%. Maximum recovery occurred during the first 3 days of a 9-day period. *Kritchevsky et al.* [508] used the Zilversmit isotope ratio method to calculate cholesterol absorption. They fed [14]C- or ^{3}H-cholesterol and injected i.v. oppositely labeled cholesterol, calculating the absorption from the ratio of oral versus injected cholesterol. Corrections were made for degradation or disappearance of sterol by using 22,23-^{3}H-sitosterol. The oral dose was 50 μCi for baboons and 25 μCi for vervet monkeys. The i.v. doses were 10 and 5 μCi, respectively. 4 male baboons, *Papio ursinus,* and 6 vervet monkeys, *Cercopitheus aetiops pygerethrus,* were used. Cholesterol absorption was 26.3 and 26.9%, respectively, for the two species. By the Grundy fecal method, the corresponding figures were 22.2 and 29.1%, respectively. The uncorrected cholesterol absorption was 90 and 92%, and the sitosterol degradation was 64 and 69%, respectively. The correction for sterol degradation must be determined for each species and, preferably, for each experiment.

All the other studies were made with rats. *Siperstein et al.* [773] proved that bile is an obligatory requirement for the transport of cholesterol from the intestine into the lymph. Upon this disclosure, the analysis of thoracic lymph became a favored method for the study of sterol absorption. *Hanahan and Wakil* [354] reported that 2–5% of ingested ^{14}C-ergosterol in 0.5 ml corn oil appeared in the thoracic duct of rats within 6 h. None appeared in the bile. *Hernandez and Chaikoff* [380] and *Hernandez et al.* [381, 382] gave 4–100 mg 4-^{14}C-cholesterol in cottonseed oil ± 4–100 mg soy sterols or $\pm$ β-sitosterol by stomach tube to 20 rats and collected thoracic lymph for 24 h. When fed alone, about 30% of the cholesterol was recovered, but only 0.5% was recovered when fed together with soy sterols or β-sitosterol. In their first experiment, less esterified than non-esterified cholesterol was found in the lymph and this led to the conclusion that β-sitosterol interferes with cholesterol esterification. However, no differences were noted in a second experiment. *Gould* [326–328], who studied sitosterol absorption in rats and in man, found that in rats fed either mixed sitosterols or pure β-sitosterol (90%) the absorption of sitosterol was about one tenth that of cholesterol. Of the ingested sitosterol, 1% was found in the blood, 1–2% in the liver, 2% in the digestive tract, and 4–6% in the carcass. Sitosterol was excreted in the bile. It had no effect on hepatic cholesterol synthesis, as measured by the incorporation of ^{14}C-acetate. Years later, *Shefer et al.* [759] alerted us to the fact that diurnal differences of 3-hydroxy-3β-methylglutaric acid-coenzyme A reductase (HMG-CoA reductase) limit the conclusions that can be drawn from studies of acetate incorporation.

Swell and co-workers made intensive studies of the sterol metabolism of rats – before and during the isotope era. Six [838, 841, 845–848] of their many essays are summarized here. Administration of 25 mg (2.04 µCi) of ^{3}H-β-sitosterol to rats fed a cholesterol-free diet with 70 mg sodium taurocholate, 50 mg egg albumin, and 75 mg oleic acid significantly enhanced fecal excretion of cholesterol and related sterols. The calculated loss of sitosterol was either 32 or 53% of the injected quantity. Rats with lymph fistula received 44 mg ^{3}H-β-sitosterol. Analysis at 6, 24, and 48 h revealed that fecal excretion of sterols doubled. Uptake of ^{3}H by the intestinal wall was considerable – 48%/48 h. No ^{3}H appeared in the lymph. In other studies, 48 mg ^{14}C-phytosterol lowered the amount of lymph cholesterol. When ^{14}C-phytosterols were fed alone, only 0.34% of the ingested amount was found in the lymph after 6 h and only 2.1% after 24 h. It was thought that plant sterols are not absorbed via lymph but by another route, and that up to 33% of ingested phytosterols were metabolized in the intestines to substances which are not precipitated by digitonin. Labeled phytosterol glucoside was poorly absorbed, compared to cholesterol glucoside. *Berthold-Godefroy and Wolff* [85, 86] fed 9 non-fasting rats by stomach tube a mix of 0.25 ml cottonseed oil with 4 mg cholesterol-4-^{14}C and 4 mg of unlabeled cholesterol or 25 mg sitosterol. Analysis of the feces and the intestinal content revealed that sitosterol decreased the cholesterol absorption from the control level of 70% to a level of 56%.

Different conclusions were reached by *Gerson et al.* [297, 298], who kept rats on a low-fat diet for 3 months. One group of rats was injected i.p. with 5 mg of emulsified β-sitosterol on the last 25 days of a 3-month period. During this time, food consumption decreased from 13 to 7 g/day/rat and adipose tissue diminished. 1 h before the rats were killed they received i.p. CH_3-^{14}COO Na. The CO_2 expiration rate increased by 7 with a 14% increment in the ^{14}C content of CO_2, compared to controls. No increase in cholesterol biosynthesis was shown. Sitosterol seemed to enhance the rate of degradation with a concomitant increase in CO_2 expiration and of the ^{14}C content of CO_2. The specific activity was lower on a fat-rich diet than on a low-fat diet, $^{14}CO_2$ expiration was higher on the high-fat diet. In the presence of dietary fat, the sitosterol effect on plasma and liver cholesterol increased. Sitosterol depressed plasma cholesterol by 15 mg/fat-fed rat, compared to its effect in rats on a low-fat diet. Yet, the authors concluded that sitosterol does not block the absorption of cholesterol.

Dunham et al. [240] utilized the combined feeding of ^{14}C-cholesterol and recrystallized ^3H-β-sitosterol to 43 rats with cannulized thoracic lymph duct. Sitosterol consistently lowered and delayed the absorption of cholesterol. Its absorption was about 10% of the absorption of cholesterol. Lymph cholesterol was esterified to 70%, lymph β-sitosterol to 25%. Excess sitosterol was required to counteract the 'dilution' with autogenous cholesterol. The results were striking when 30 mg of β-sitosterol was given. This amount was roughly equal to the limit of sitosterol solubility in corn oil in the presence of cholesterol. Concurrent studies using double-labeled sterols were also conducted by *Sodhi et al.* [781]. Rats received [4-^{14}C]-cholesterol and [1,2-^3H]-β-sitosterol in oil within gelatin capsules (S:C = 6:6) and a second capsule with 500 mg carmine red dye as fecal marker. 6 h after the 'meal', four segments of the small bowel and two of the large bowel were removed for assay. The lumen-mucosa-lumen cycle (in 6 rats) was short and there was no difference in percent excretion. By the ratio method, the values were $3 \pm 7\%$ greater. The difference between the two methods was from −7.3 to +14%. In the absence of carrier cholesterol, the absorption was over 36%. In the presence of the carrier it decreased, and reached 30% on 6 mg cholesterol ingestion.

Zilversmit [955, 956] compared the dual isotope method and the single double-label isotope method administering [^{14}C]- and [^3H]-cholesterol intravenously. By the dual method, cholesterol absorption was 45%, with the single method 49%. Next, he used i.v. and orally administered

4-^{14}C-β-sitosterol and the [^{14}C]- and [^{3}H]-cholesterol. In 5 rats, ^{3}H/^{14}C of the plasma was 32.7–48.6 within 24 h, 32.8–51.6 within 48 h, and 34.6–49.0 in 96 h. In the feces, neutral steroid ^{3}H was 27.7–34.2 after 1 day, 11.5–22.4 after 2 days, 0.8–3.0 after 3 days, and 0.5–1.6 after 4 days. The total fecal ^{3}H for the 4 days was 36.8–57.1. This pointed to the need for corrections: (1) for labeled cholesterol which was absorbed and then excreted into the lumen, thus lowering the value of uncorrected non-absorbed labeled cholesterol, and (2) for the loss of sterol due to bacterial degradation, thus raising the value for non-absorbed cholesterol.

Bile-fistula rats were used by *Subbiah* [801] and *Subbiah and Kuksis* [813, 814]. Rats were injected i.v. with lecithin emulsion containing 4-^{14}C-β-sitosterol and bile was collected for 5 days. Sterols were separated on the basis of hydroxyl groups by TLC on silica gel and the monohydroxysterols were resolved on the basis of unsaturation by TLC on silica gel with AgNO$_3$. The bile contained 38% of sterols in the neutral fraction and 60% in the acid fraction. Mono-, di- and trihydroxy acids were present. In the liver, radioactive distribution was 85.4% β-sitosterol, 11% β-sitostanol, 3% dienols, and 0.6% trienols. In other experiments, 5 μCi of labeled sterols (1 μCi/mg) in 0.5 ml ethanol-diethylether 1:1 v/v plus 5 mg purified egg lecithin and 5 ml 0.9% NaCl solution sonicated in ice at an output of 40 W, were used. The final mixture for i.v. injection contained 1 μCi/mg sterol/1 ml solution. Sitosterol moved rapidly from plasma to liver to bile. The total radioactivity per plasma volume equalled 3.5% of body weight. The bile contained 10% of the injected activity in the first 24 h, 20% in 2 days, and 25% in 3 days. Over a 5-day period the bile fistula yielded 24% sitosterol, 60% as acid and 40% as neutral steroid, and 3% cholesterol, 85% as acid and 15% as neutral steroid. After simultaneous injection of ^{14}C-cholesterol and ^{3}H-β-sitosterol, cholesterol cleared more rapidly from blood and more of it appeared in liver and bile. The liver cleared sitosterol less readily than cholesterol. Plasma and adrenal sitosteryl esters were 40–50% lower than cholesteryl esters, regardless of whether the two sterols were injected together or separately. Sitosterol seemed to associate preferentially with PUFA in the liver.

Siegfried and Hyde [768] cannulated the bile duct of rats. 2 weeks after ingestion of 1 mg cholesterol-4-^{14}C or of 0.2 mg β-sitosterol-3-^{14}C, the bile contained 80% of cholesterol and 42% of sitosterol. After alkaline hydrolysis and extraction, the neutral fraction contained 4% of cholesterol and 80% of sitosterol-^{14}C, the acid fraction 78 and 5%, and the aqueous fraction 9 and 11%, respectively. Hepatic conversion of β-sitosterol to bile acid end products was considerably less than for cholesterol. The differences in the results of experiments with rats can be ascribed to the many modifications in experimental design.

b) Man

In contrast to the confusing results with animals, clinicians applied the various isotope methods successfully to sterol balance studies in man.

The first reports on absorption of tritium-labeled β-sitosterol in man came from *Gould* [326–328] and *Gould et al.* [330]. Of a 400-mg oral dose of sitosterol, 0.1–1% appeared in the blood and liver. The absorption rate was later fixed as 10% of that for cholesterol. 50 healthy men and 5 terminal patients received a single 50-mg dose of ^{3}H-β-sitosterol. Of that, only 0.015 mg/dl was recovered from plasma and erythrocytes. Of a single dose of 258 mg (6.7 µCi) of ^{3}H-β-sitosterol, given to a patient 5 days before his demise, 0.5% was present in the viscera (except the intestines). On a dose of 1 µCi, the peak activity for plasma-free sterol was 9,200 dpm/mg (0.35 µCi), which was 0.295 µCi/mg lower than upon intake of 0.1 µCi of ^{3}H-cholesterol. The postmortem blood of a patient who died 38 days after he had received 5.7 µCi (258 mg) of labeled sitosterol contained 0.0087 µg/mg; the liver had 0.0096, the bile 0.0326, the intact aorta had 0.0033, the atherosclerotic aorta had 0.0016, the pulmonary artery 0.0045, and atherosclerotic coronary artery tissue had 0.0045 µg/mg. In contrast to these observations, *Friedman et al.* [284] stated that – in conformity to their experience with rats – ingestion of sitosterol (dose not given) for 6 months had no effect on intestinal cholesterol absorption in 6 patients with coronary disease.

Borgström [137] introduced the one-oral-dose method for sterol balance studies, using 4-^{14}C cholesterol plus 22,23-^{3}H-β-sitosterol. On ingestion of 150 mg β-sitosterol dispersed in the butter of a standard breakfast given to 20 subjects, 90% of the oral dose was recovered in feces. When 70% sitosterol plus 30% campesterol was fed, with or without cholesterol, the cholesterol excretion was 60–80%, regardless of the amounts (150–1,810 mg) of ingested cholesterol. Absorption of cholesterol is, according to *Borgström,* limited and does not exceed 200–300 mg/day, even when larger doses are taken. This is in accord with observations made by *Beveridge et al.* [113]. 28 persons were given 150–1,910 mg cholesterol in 50 g butter (1 µCi 4-^{14}C-cholesterol/50 mg) and ^{3}H-β-sitosterol was added as an external standard to feces. Next, 4 persons were given 150 mg of each of the two sterols. Cholesterol absorption was 96%, sitosterol absorption was 8% (i.e. 3–6% of the oral 150-mg dose). In 1 man, however, only 40% of the sitosterol was excreted and cholesterol excretion was also lower than in all the others.

Grundy et al. [346] used β-sitosterol as an internal standard. Although it should be 'totally' recovered, only 65% of the oral dose was found in feces in 'many' patients. This was ascribed to bacterial degradation of neutral 3β-OH-Δ^5-sterols to products which are not recognizable as steroids on fecal analysis. In man, on a low dose, 37% of dietary β-sitosterol are degraded, on a high dose only 23%. In the feces of 2 patients in whom cholesterol-^{14}C and β-sitosterol-^{3}H were instilled into the terminal ileum (from which neither sterol is absorbed), 25% less of either sterols could be recovered. In 4 patients who received ^{14}C-cholesterol daily by mouth until an isotope steady state was reached, the loss was 28–50%. However, labeled bile acids were recovered to 100%. *Blomstrand and Ahrens* [131] studied the absorption of dietary cholesterol-^{14}C (2.4 µCi/mg) in a patient with chyluria. Simultaneous ingestion of 8.3 mg cholesterol and 5 g β-sitosterol (based on a circulatory extrahepatic pool of 1–2 g) reduced cholesterol absorption by over 50%.

Quintão et al. [678] evaluated in 12 patients four methods by which cholesterol absorption had been measured: (1) cholesterol-4-^{14}C in one i.v. dose; (2) daily oral cholesterol ingestion; (3) isotope steady-state method, and (4) Bergström's one-dose oral method with 4-^{14}C-cholesterol plus 22,23-^{3}H-β-sitosterol. The disadvantages of the first three methods lie in the contribution of unlabeled cholesterol to the intestinal contents plus addition of synthesized cholesterol. The fourth method requires stool collections for up to 8 days, i.e. as long as there is activity. Essentially non-absorbable sitosterol serves as internal standard. Cholesterol absorption is calculated as the loss of cholesterol relative to sitosterol during intestinal transit: it is the difference between the test dose and the recovery of fecal neutral steroids. Prior to the test, the mucosa will have reached an equilibrium for non-radioactive exogenous and endogenous cholesterol and also for sitosterol. This equilibrium would not be upset by a single dose with a negligible mass. The increase of the ^{14}C/^{3}H ratio in progressive samples indicates a 'back-exchange' in the intestinal mucosa. *Ahrens* [7] commented further on the methods by which sterol balance can be studied. In the 'steady state' as much cholesterol is excreted as was ingested and recovery of sitosterol is quantitative. It amounts to 95% in most patients. In some patients it is only 60% but since recovery of cholesterol is also incomplete (a percent match) the loss cannot be explained by a prolonged stay of the sterols in the colon. Thus, loss of radioactive fecal sterols must be corrected by the percent of recovered sitosterol. Inadequate bowel evacuation must be taken into account. In the single-isotope-dose method, instillation of the two labeled sterols in the proximal colon should result in complete recovery.

Kudchodkar et al. [512, 513] measured the absorption of dietary sterols in 10 patients with hyperlipidemia. They used the balance method. 9 patients equilibrated on their controlled habitual diets. 3- to 4-day fecal pools were collected for analysis. There was good correlation between intake, absorption, and excretion. On intake of 564–1,212 mg cholesterol/day, 149–529 mg/day (37.5%) were absorbed, and on intake of 171–450 mg β-sitosterol/day, 6% was absorbed. Minor sample variations were –12 ± 6%/subject. Mean recovery of β-sitosterol was 92 ± 5%.

Using the ratio method in a study of 20 men, *Sodhi et al.* [781] gave them cholesterol-4-^{14}C plus β-sitosterol-1,2-^{3}H in corn oil in a gelatin capsule, and 500 mg carmine in a second capsule, as they used for rats. The average absorption value, corrected for β-sitosterol, was 2% higher than when measured by Borgström's method. The correlation for the two methods was r = 0.89. The range of cholesterol absorption was 43–71% for the ratio method, 70.5% for Borgström's method. This was explained by the absorption of ^{14}C; while some of it is transported by lymph, most reappears in the intestinal lumen. Meanwhile, the unabsorbed portion has moved down the intestine, together with the cholesterol which had been synthesized by the intestinal wall and discharged into the lumen. Released cholesterol is apparently not as readily absorbed as dietary cholesterol.

Cevallos et al. [162] used the total balance method and the isotope dilution technique. They monitored the plasma ^{14}C-cholesterol decay curve after i.v. injection of 4-^{14}C-cholesterol for a 55- to 90-day period. Pooled Cr$_2$O$_3$ and β-sitosterol were added for correction of fecal flow and sterol losses. In 2 normal men and in 3 hyperlipoproteinemic men, cholesterol production was 1,408 ± 845 mg/day and excretion was 1,377- ± 914 mg/day.

β-Sitosterol has become widely accepted as the best internal marker since it has all the molecular characteristics of cholesterol but is absorbed only minimally. *Stamnes et al.* [790] used it, together with Cr$_2$O$_3$, as marker to prove that a 3-week exchange of 1–2 g sucrose for starch/kg body weight in solid food did not cause a consistent change in serum lipids or cholesterol balance in 3 carbohydrate-sensitive persons and in 1 normolipemic man. *Briones et al.* [145] used Cr_2O_3 and β-sitosterol as fecal markers in a study of excretion of bile acids and neutral sterols in 7 normal subjects, in 4 with hyperlipoproteinemia of heterozygous type IIa, and in 1 with homozygous type IIa. Bile acid excretion was 353 mg/dl in the controls, 204 mg/dl (p < 0.05) in heterozygous IIa, and 420 mg in homozygous IIa hyperlipoproteinemia. Neutral sterols in bile were 173 mg/dl in controls and 483 mg/dl in heterozygous IIa hyperlipoproteinemia. Neutral sterols in bile were 173 mg/dl in controls, 483 mg/dl in heterozygous IIa, and 725 mg/dl in homozygous IIa. Synthesis of bile acids plus neutral sterols (minus cholesterol intake) was 833 mg/dl in controls, 534 mg/dl in heterozygous type IIa, and 947 mg/dl in homozygous type IIa.

Kottke and Subbiah [494, 495] gave to 8, later 9, hyperlipemic patients 1 g sitosterol (Cytellin) and 1 g Cr_2O_3 daily as flow markers. Fecal neutral sterol recovery was 92%/day for 9 days, measured by GLC and TLC. Recovery of the two markers was comparable, the mean difference was 11.4%. In 1 patient, recovery of Cr_2O_3 was higher while in the other 7, that of sitosterol was higher. In all patients, recovery of sitosterol, when ingested alone, was greater than that on campesterol-cholesterol intake: 98% against 86%. *Subbiah et al.* [815] gave 10 patients with hyperlipoproteinemia a solid diet plus 1 g/day of plant sterols (sito- and campesterol) and 1.5 g/day chromium sesquioxide. There was a 20% difference in the recovery of the markers: 68% for Cr_2O_5, 89% for phytosterols. *Kottke et al.* [493] followed 17 patients with different types of hyperlipoproteinemia. Cholesterol synthesis was increased in patients with type IV and V, with a significant correlation (r = 0.05) between neutral sterol excretion and cholesterol ingestion. Bile acids were higher in types IV and V, lower in type II. *Subbiah* [803] concluded from his steady-state balance studies that it is not sufficient to measure β-sitosterol alone, since preparations taken orally contain other sterols, especially campesterol. Balance studies must take into consideration the excretion of total steroids, neutral sterols, and acidic steroids. He summarized his conclusions in the following formula: cholesterol excretion/day = cholesterol + metabolites × β-sitosterol/sample + [β-sitosterol intake/day].

Bhattacharyya et al. [126] reported that 2 men whose diet contained 700–1,000 mg cholesterol/day excreted 103 and 81 mg/dl of it via the skin. 4 men, without dietary cholesterol, excreted via the skin 98, 85, 58, and 59 mg/dl cholesterol. 6 normocholesterolemic subjects (average level 193 mg/dl, range 214–154 mg/dl) and 5 with hypercholesterolemia (average level 463 mg/dl, range 295–864 mg/dl) received 4-^{14}C-cholesterol or

[1α³H]-cholesterol plus 400 mg plant sterols. After 24 h, the skin was wiped with acetone and their clothing was subjected to chloroform-methanol extraction. By TLC and GLC, the relative retention time of β-sitosterol was 99.6% against 5α-cholestane. Of the sitosterol, 90% was collected from the clothing and 10% from the acetone skin wash. Over 6 days, those with normal plasma cholesterol excreted 129 mg cholesterol (84% as ester), 4.8 mg plant sterols, and 1.8 mg other steroids. The specific radioactivity of the skin surface rose gradually after i.v. injection and within 4–5 weeks equalled the plasma activity. It remained high for up to 10 weeks. In all subjects, 7% of the total sterols was the same as in the feces, i.e. 0.33:0.41:7.39 mg/24 h.

Nikkari et al. [614] criticized this procedure since no analysis had been made of the hair and of soiled pajamas of patients, and because boiling acetone was used for sterol extraction. They felt that extraction should have been performed on the water from the subjets' shower at the end of 24 h, and that chloroform-methanol azeotrope 99:12 v/v should be used for extraction. They gave 10 subjects seven different diets with 25 mg phytosterols added (90% β-sito-, 10% campe- and stigmasterol) in capsules per each 100 g of the diet formula. Ingestion of 5.5 g Cytellin (i.e. 3.4 g β-sitosterol) daily for 1 month did not alter the amount of β-sitosterol in the skin folds. No β-sitosterol was present in the skin of subjects on a corn oil diet. The cholesterol extraction did not vary either, averaging 88 mg/dl for 8 normocholesterolemic subjects. According to *Bhattacharyya et al.* [126] 6 men with normal plasma cholesterol excreted 82 mg/dl.

3. The Post-Isotope Era

The heading of this section has been explained in the introduction to the discussion of sterol absorption. This era is characterized by the utilization of precise chromatographic methods (GLC, TLC, etc.) which allow determination of sterols in amounts as small as 0.01 mg/dl. The statement by *Weizel* [911] that 'the question concerning absorption of β-sitosterol has been solved with the use of isotopes' was somewhat too optimistic. There is no strict dividing line between multiple approaches: chromatographic methods are often combined with isotope studies, as mentioned before.

a) Animals

Dogs with cannulated thoracic ducts were observed by *Kuksis and Huang* [515]. They were fed 5–10% solutions of a plant sterol mixture in oleic acid or in stripped corn oil. Lymph flow of 3,050 ml/h was collected over 20 h and 2-hour collections were analyzed. Total sterols were measured by a digitonide-anthrone procedure; plant sterol:cholesterol ratios were determined by GC of sterol acetates. Maximum absorption of all sterols and of their esterified portions (50%) occurred between 4 and 8 h after feeding. Total lymph sterols rose from the fasting level of 3.7 mg/dl to a postprandial level of about 50 mg/dl. Apparently, the authors were unaware of the fact that γ-sitosterol is a mixture which contains a considerable amount of cholesterol when they reported that γ-sitosterol was absorbed 4.5 times as readily as β-sitosterol and that on feeding soybean sterol mixture which contained 40% of γ-sitosterol and 50% of β-sitosterol, 75% of the absorbed plant sterols was the γ-isomer.

Rats were fed a stock diet containing one of five differents supplements by *Cohen et al.* [180]. The conversion rates of cholesterol were assayed by GLC and TLC. The daily excretion of coprostanol was 35% (or 5 mg) on a 5% corn oil-supplemented diet, 32% on sodium taurodeoxycholate addition, only 8% on 0.5% sodium taurocholate; it was 31.5% on 1.2% cholesterol and 23% on 0.8% β-sitosterol supplements. The differences were ascribed to the effect of bacterial degradation of cholesterol.

Pertinent contributions were made by *Subbiah* [805], *Subbiah and Kottke* [806, 807], and *Subbiah et al.* [809]. They compared atherosclerosis-susceptible White Carneau pigeons and atherosclerosis-resistant Show Racer pigeons. Both breeds received feed containing a mixture of 19.4% cholesterol, 22.5% campesterol, 50.8% β-sitosterol and 7.3% stigmasterol. The fecal $^5Δ$-sterols mirrored the composition of the diet. The 5α-stanols were derived from 16.8% cholesterol, 27.4% campesterol, and 55.8% stigmasterol. Fecal bile acids were chenodeoxycholic acid, lithocholic acid, 7-ketolithocholic acid, traces of cholic and of deoxycholic acids. White Carneau pigeons, 4–6 years old, were fed a diet containing 74.9% β-sitosterol, 7% campesterol and 8.5% stigmasterol. Some of the birds had bile fistulae. Intestinal segments of the birds were washed free of contents with saline and 1 mM sodium taurodeoxycholate. The retention times on GLC, relative to cholestane, for trimethyl silyl esters on 3.8% W-98 columns were 4.05 for β-sitosterol, 3.16 for campesterol, and 3.47 for stigmasterol. The proximal intestine contained 98% cholesterol and 2% plant sterols. The proportions of the three plant sterols was the same as in the diet. The fecal ratio of campesterol to β-sitosterol was 2.34, which was ten times more than in the diet. This indicated preferential uptake of campesterol. In the stomach the ratio of these sterols was 1.23, in the duodenum 3.16, in the proximal intestine 2.34, and in the distal bowel 3.92. The mean plant sterol concentrations were 13.8 µg/g for the stomach,

Table XXXI. Fecal sterols and stanols of six avian species. After *Subbiah et al.* [810]

Compound	Pigeon	Rooster	Duck	Goose	Owl	Hawk
Δ^5 Sterols						
Cholesterol	+	+	+	+	+	+
Campesterol	+	+	+	+	+	+
Stigmasterol	+	+	+	+		
β-Sitosterol	+	+	+	+	+	
5α-Stanols						
Cholestanol	+	+	+	+		
Campestanol	+	+	+	+		
Stigmastenol	+					
Sitostanol	+	+	+	+		
5β-Stanols						
Coprostanol		+	+			
Coprocampestanol		+				
Coprostigmastanol		+				
Coprositostanol		+	+	+	+	
$\Delta^{5,24}$ Sterols						
Desmosterol	+			+		+
β-Ketosteroids						
Coprostanone		+				
Campestanone		+		+		+
Stigmastenone						+
β-Sitosterone		+	+	+		+
Unknown	+	+	+			

13.4 µg/g for duodenal tissue, 28.45 µg/g for proximal and 19 µg/g for distal intestine. Only traces of stigmasterol were found although in the diet it was one half of the amount of campesterol. 4 h after ingestion of 1 g sitosterol (as Cytellin, 65% β-sitosterol, 30% campesterol, 5% stigmasterol) the mean campesterol: sitosterol ratio was 1.5 in the small intestine and 4.1–4.9 in liver and bile, compared to 0.46 in the diet. The skin of 2-year old White Carneau pigeons was minced and the extract analyzed. The bands obtained on AgNO₃-impregnated silica gel chromatograms were, after trimethyl silylation elution, subjected to GLC. They found 94.2% cholesterol, 0.4% cholestanol, 0.2% campesterol plus β-sitosterol plus traces of minor sterols.

Subbiah et al. [810] compared the proportions of fecal sterols and stanols in several avian species and expressed the amounts in µg/g of feces (table XXXI).

b) Man

The first studies in which chromatographic methods were systematically applied to analysis of the blood lipids before and after ingestion of plant sterols by man were reported by *Böhle et al.* [133]. 18 subjects, with an average age of 46 years, were healthy and had normal blood lipid levels. 13 others were 53-year-olds and had hypercholesterolemia and atherosclerosis. Of all the 31, two thirds had traces of phytosterol in their serum prior to phytosterol ingestion. Small as the amounts were, they were five times higher in the control group than in the patients. The normocholesterolemic subjects had 212 mg/dl of cholesterol and 1.4 mg/dl of plant sterols, and 802 mg/dl of total lipids in the serum. For the hypercholesterolemic, the corresponding values were 256, 0.22, and 1,060 mg/dl. Peroral doses of sitosterol caused a small but universal increase in total phytosterols. 8 patients who had an altered fat metabolism received for 3 months either 18 g/day of granulated sitosterol or a vegetable oil which was rich in plant sterols or in polyenoic acids. This resulted in a marked decrease in TL and TC, without a significant increase in plasma phytosterols. It was concluded that healthy human beings can absorb plant sterols in small amounts and to various degrees.

Salen et al. [716, 717] used the simultaneous isotope turnover and balance methods. Three men who had a constant weight were given diets containing 250 mg cholesterol and 100 mg β-sitosterol (S:C = 1:2.5) per 1,000 cal. By GLC of biweekly plasma samples, cholesterol was 287, 287, and 220 mg/dl in the 3 men, and β-sitosterol 1.02, 0.5, and 0.3 mg/dl, respectively. On pulse labeling with 22,23-^{3}H-β-sitosterol i.v., the specific activity conformed to the two-pool model. Since β-sitosterol is not synthesized by man, the turnover equals the absorption. This absorption was, respectively, 12.5, 7.5, and 6.5 mg/day, i.e. 1.5–2.5% of the intake. In contrast, cholesterol absorption was, by fecal measure, 204, 307, and 307 mg/day, i.e. 45–54% of the intake. At first, it was thought that only 1% of the absorbed β-sitosterol is converted to cholic and chenodeoxycholic acids; later, the figure was corrected to 20%. Excretion of β-sitosterol is speedier than that of cholesterol. Biotransformation to bile acids is analogous for the two sterols. In 12 patients with hypercholesterolemia, plasma β-sitosterol remained constant on fixed intake of β-sitosterol. Specific activity (SA) curves after simultaneous labeling with ^{4}H-β-sitosterol and ^{14}C-cholesterol confirmed the two-pool model. The two exponential half-lives of β-sitosterol were much shorter than for cholesterol; pool sizes were much smaller. Turnover values for β-sitosterol were similar in isotope balance methods. The average plasma cholesterol was 226–300 mg/dl, the average β-sitosterol was 0.3–1.05 mg/dl. Packed red blood cells had 105–131 mg/dl cholesterol and 0.7–0.2 mg/dl β-sitosterol. A diet containing 40% calories from corn oil plus 620 mg of 22,23-^{3}H-β-sitosterol (SA 72,245 dpm/mg) was given for 83 days. Then sitosterol (SA 1,000 dpm/mg) was injected and a steady state was reached. The SA of β-sitosterol was 76,149 dpm/day, or only a 3% increment. The agreement between dietary and fecal activity was ± 5%. The authors concluded that 10 times more cholesterol than β-sitosterol is returned to the body pool through reabsorption, and that the liver cells are capable of distinguishing between the two sterols whereas the enzymes which convert sterols to bile acids are not capable of making such a distinction. The possibility that β-sitosterol was dealkylated to cholesterol prior to its incorporation into bile, alone or simultaneously with side chain oxidation during bile synthesis, was considered.

Recently, *Oster et al.* [638, 639] observed 25 patients who had a high plasma level of LDL cholesterol, 7 patients with type IIb hyperlipoproteinemia with high TG, and 8 patients with type IIa hyperlipoproteinemia, for 16 weeks. The maximum plasma sitosterol level in any patient was 1.2 mg/dl. In only 1 patient did the level exceed 0.001 mg/dl prior to ingestion of sitosterol. 7% of samples of normal plasma contained sitosterol whereas this sterol was found in 31% of samples of patients being given 15 g of β-sitosterol daily. In 15 adults who received 24 g β-sitosterol/day plasma phytosterols always averaged less than 0.3% of their plasma cholesterol level.

Grundy and Mok [347] reported on the effect of low doses of phytosterols on the absorption of cholesterol. They assumed that β-sitosterol interferes with the absorption of exo- and of endogenous cholesterol, that it increases fecal neutral steroids, and that it affects the drain on the liver cholesterol pool. A 3-g/day dose of pure β-sitosterol powder – a dose considered adequate to lower plasma cholesterol – resulted in a plasma β-sitosterol level of less than 1 mg/dl. Sitosterol absorption was less than 5% and its excretion was rapid. 10 patients with normal or elevated plasma cholesterol ingested for 3 months 40% of their calories as fat (lard); to the diet of 3 patients, 700 mg of cholesterol was added. Cytellin (90% β-sitosterol) was added for cholesterol balance studies as internal standard and Cr_2O_3 was added to correct acidic steroids for the variations in fecal flow. Four 5-day pooled fecal samples were analyzed by GLC. Orally, ^{14}C-cholesterol was given to measure the daily absorption. The authors present a novel technic based on intestinal intubation and on the aspiration of samples, calculating that the net cholesterol absorption in mg/h equals the TC in the inflow minus the TC in the outflow (mg/h). The influx C:S ratio was measured at: tube No. 1, ampulla outlet; tube No. 2, 10 cm below No. 1, and × sitosterol input at No. 1. The outflow C:S ratio was measured at: tube No. 3, 100 cm below No. 2, and × sitosterol input at No. 1. Plasma cholesterol had a downward trend, in spite of fluctuations. Endogenous cholesterol may have exceeded the exogenous cholesterol supply. On the basis of this study, the authors recommended sitosterol as a hypocholesterolemic agent, to be used alone or in combination with drugs.

Lees and Lees [529] conducted a rather definitive study on the degree of absorption of β-sitosterol and campesterol. On ingestion of 6 g of liquified, spray-dried dispersible sitosterol powder (93% β-sitosterol), the plasma β-sitosterol averaged 0.03–2.45 mg/dl for 8 persons. On ingestion of 3 g/day of this preparation by 19 patients, the plasma β-sitosterol averaged 0.47–2.07 mg/dl. In 18 patients who ingested 18 g of suspension (with 60–65% β-sitosterol) daily, the absorption of β-sitosterol was 0.37–0.75 mg/dl. This came up to 10% of the dose taken. The average plasma levels were 0.48%/dl for β-sitosterol, 16.8 mg/dl for campesterol, and 240 mg/dl for cholesterol.

In summary, one can say that phytosterols are or can be absorbed by man, though apparently not by all. They are absorbed better by normocholesterolemic subjects than by hypercholesterolemic ones: plasma phytosterol levels are up to five to six times higher in normo- than in hypercholesterolemic subjects. The average levels are 0.22–1.4 mg/dl for the first and 0.2–0.89 mg/dl for the second type of patient. The maximum amount of plasma phytosterols found was 4.7 mg/dl. To be sure, there are exceptions to every rule, and one should bear in mind the reports on patients with hypersitosterolemia.

The repeated disclosure that campesterol is absorbed to a considerably greater extent than β-sitosterol, in atherosclerosis-prone White Carneau pigeons and in atherosclerosis-prone man, is important.

B. Mechanisms of Cholesterol Reduction by Sitosterol

Multiple theories have been advanced concerning the mechanism by which β-sitosterol functions as a plasma cholesterol depressant. In 1961,

Chiu [167] listed and reviewed these theories. However, the list has grown considerably since then.

We have to consider the possibility that more than one mechanism could be involved, simultaneously, or separately. Many of the theories which have been put forward have already been discarded, others have yet to be proven right or wrong. No concept has as yet found universal acceptance.

1. Mixed Crystals

The first attempt to explain the interference of β-sitosterol with the absorption of cholesterol from the intestinal lumen was made be *Pollak* [660]. In rather unsophisticated *in vitro* experiments cholesterol and β-sitosterol were purified by repeated recrystallization, dissolved separately, mixed in various proportions, and recrystallized again. The formation of 1:1 mixed crystals was observed. It was suggested that such crystals would be too large to penetrate the membrane of intestinal mucosal cells.

Some data on mixed crystals of sterols were, however, already available. *Dam* [210], in 1934, conducted a single trial: in a man who received 3.45 g of dihydrocholesterol in butter over 12 meals in 5 days, inhibition of the absorption of cholesterol was ascribed to the formation of non-separable crystals. *Bissett and Cook* [129] stated that after ingestion of 10 g of mixed sitosterols by 3 men for 5 days mixed crystals were present in the feces. *Fujimoto and Jacobson* [288], describing in detail the preparation of β-sitosterol-4-^{14}C from tall oil, found that structurally related plant sterols formed mixed crystals. This situation hampered the purification of β-sitosterol. *Murakami et al.* [601] also had difficulty in isolating β-sitosterol because of the presence of mixed crystals of two sterols in plants: in root bark of *Aralia elata* (Mis) Seeman, one part of β-sitosterol was bound to two parts of stigmasterol. In three other plants, *Melia azedarach* var. *japonica, Vitex cannibifolia,* and *Pinellia ternata,* crystals were composed of one part β-sitosterol and two parts campesterol. All the reports on 'γ-sitosterol', cited earlier in this text, refer to mixed crystals: in the case of γ-sitosterol from soy beans, there were crystals of β-sitosterol plus campesterol, and in the case of γ-sitosterol from toad secretions, the crystals were a mixture of cholesterol, β-sitosterol and campesterol.

Davis [216] made an extensive physicochemical study on the solubility of sterols and on the relationship of mixed crystals. In water, the solubility of unsaponifiables is less than 0.1 μg/l. In methanol, at 25 °C, the solubility of cholesterol crystals is 6 mg/ml and that of β-sitosterol is

2 mg/ml. In a 1:1 mixture, the solubility of the two sterols in combination is 2.4 mg/dl. Thus, β-sitosterol considerably reduced the solubility of cholesterol. The different degree of solubility was explained by the lattice energy of crystals. Next, he studied the solubility in sodium oleate: sitosterol was less dispersable than cholesterol. In mixture, new crystals were formed which, on X-ray diffraction (XD), resembled crystals of epicholesterol-β-cholestanol. Various amounts of added β-sitosterol, β-cholestanol, or stigmasterol decreased the solubility of cholesterol. *Shipley* [761] elaborated on these findings. He stated that there seem to be certain prerequisites for the *in vivo* formation of mixed cholesterol-sitosterol crystals: a good mixture of β-sitosterol with the cholesterol-containing food, a surplus of β-sitosterol, fine dispersion of the β-sitosterol, and a diet which contains fat and FAs.

This mixed-crystal concept was criticized because all the experiments had been made *in vitro*. *Hudson et al.* [409] countered the argument. They used sodium deoxycholate for *in vitro* studies to give a baseline: the solubility of mixed crystals was further reduced in aqueous sodium deoxycholate or sodium oleate. Next, rabbits were fed a purified diet without fiber but with finely powdered 2% cholesterol and 8% β-sitosterol blended into the diet (S:C = 4:1). 2–3 weeks later, the rabbits were killed, the small intestine was washed, divided into two equal segments which were split and washed again. Large particles were removed initially by mild centrifugation, then by vigorous centrifugation. Fine particles with 1:1 mixed crystals were obtained and identified by XD. The solubility of these crystals was the same as that of β-sitosterol, namely one third of the solubility of cholesterol.

Wright [943–945] and *Wright and Presberg* [946, 947] studied the formation of clathrates. They found that this depends on the initial cholesterol concentration which must be in excess of one half of the saturation of cholesterol in oils. They denied that sterols compete for solubility sites, although they did think that they compete for solubility in TGs.

2. Adsorption: Aggregation

Although *Wright* [943–945] had concluded that adsorption of cholesterol on the phytosterol surface is less likely to occur than mixed-crystal formation, he tested *in vitro* adsorption of cholesterol from TG solutions and found no adsorption to Norite but some to Permutit. The effects of various compounds, including β-sitosterol, stigmasterol, lanosterol, and desmosterol, were not uniform. Multiple-layer adsorption occurred in 'concave upwards' isotherms. With a positive effect, such as induced by nicotinic acid or by caffein, fewer adsorption sites would be left for cholesterol. To what extent such *in vitro* studies could have validity *in vivo* is debatable.

Haberland and Reynolds [353] contributed a study on self-aggregation of cholesterol in water. This occurred at a critical concentration of $10^{-9}M$ and was independent of pH (5 or 10.7) and of temperature (0–50 °C). Maximum solubility of cholesterol was $4 \times 10^{-6}M$. Sterol aggregates had a partial specific volume of 0.947 mg/g (compared to 1.095 for an unassociated monomer). This was indicative of a strong interaction between monomers in the aggregation form. Density gradient configuration gave a MW of 30,000–200,000 daltons. Expansion of these studies seems desirable, using various phytosterols combined with or substituted for cholesterol.

3. Solubility of Cholesterol: Micelles

In 1955, *Ivy et al.* [417] wrote that the effect of soya sterols (containing 93% β-sitosterol) on repression of serum cholesterol is not due to the relative solubility of sterols in oleic acid or in corn oil, but rather to competition for total capacity of the sterol absorptive mechanism. *Wilkens* [924] and *Wilkens et al.* [926] were more specific. Cholesterol solubility was studied *in vitro* by mixing 1.5 g powdered cholesterol into 20 g of several oils, then examining the supernatant fluid after 20 h incubation at 37 °C. Cholesterol solubility, expressed in g/dl, was 4.33 in coconut oil, 3.96 in butter, 3.71 in beef tallow, 3.5 in olive oil, 3.48–3.46 in sunflower seed oil, 3.29 in arachis oil, 3.22 in cottonseed oil, 3.18 in cod liver oil, 3.14 in soy bean oil, 3.13 in corn oil, and 3.01 in pilchard oil. These results could explain the hypercholesterolemic effect of coconut oil and of butter. The authors also correlated the hypocholesterolemic effect of β-sitosterol with its reduction of the solubility of cholesterol: in coconut oil, the reduction was from 4.32 down to 0.68 g/dl. Addition of a synthetic unsaponifiable matter which simulated the composition of many unsaturated oils (β-sitosterol, α-tocopherol, β-carotene, squalene) decreased cholesterol solubility from 4.32 to 3.81 g/dl. *Grande and Wada* [336] reported the *in vitro* solubility of cholesterol at 37 °C as 3.22 g/dl in peanut oil, as 5.01 g/dl in coconut oil. In coconut oil plus 3% sitosterol the solubility of cholesterol decreased to 3.08 g/dl, i.e. by 40%. However, the authors thought that this phenomenon correlated positively with SFA and inversely with the iodine number of oils.

Barton and Glover [53] reported that cholesterol is dispersed in bile as a micellar complex with phospholipid and the conjugated dihydroxycholanic acid in a 1:4:7 ratio. With 3–10 mM taurodeoxycholic acid (the same amount as present in the intestines), cholesterol did not disperse. Lecithin did disperse cholesterol at a concentration of 0.4 mM of the bile salt. Ergosterol did not form micelles readily with lecithin and then only in a 1:18 proportion. This could explain the poor resorption of ergosterol.

Simmons et al. [771] perfused the upper jejunum of patients using a micellar solution of cholesterol-^{14}C, non-absorbable bile salts, and (to 93%) absorbable MGs. By radioactivity analysis of the cholesterol, 73% was absorbed, but by chemical analysis, only 46%. The SA of cholesterol fell during passage through the intestinal loop due to continued addition of non-labeled endogenous cholesterol. For 50 jejunal loops, in 35 samples, the average balance, in micrograms, was: for plant sterols, input:output = 0.46:0.75; for perfused cholesterol, input:output = 11.9:3.3; and for endogenous cholesterol, input:output = 6.1:5.0. The effect was attributed to polar lipids, the pH, the FA:bile salt ratio, the solubility effect of MG:FA micelles as factors which influence the absorption of cholesterol.

A series of ten important essays, which will be reviewed together, stems from Scandinavian investigators. Chronologically, two essays are by *Hofmann and Borgström* [395, 396] (1962, 1964), two by *Feldman and Borgström* [271, 272] (1966), two by *Borgström* [135, 136] (1967, 1968), one each by *Sylvén and Borgström* [851] (1969) and by *Sylvén and Nordström* [852] (1970), and lastly, two by *Sylvén* [849, 850] (1970, 1971).

In vivo, MGs and FAs, liberated through lipolysis, form mixed micelles with bile acids. The latter act as a detergent solution during digestion, well above the micellar concentration. Above a critical concentration micelles are formed by bile salts. All micellar solutions are water-soluble. Describing their methods, the authors obtained after an 18-hour centrifugation of heated intestinal content at 100,000 G (37 °C) a transparent aqueous micellar fraction with FAs and diglycerides, and an oil phase with more FAs and less diglycerides and TGs. They concluded that during digestion fats are selectively partitioned between the aqueous and the oil phase and that absorption takes place from the micellar phase.

Application of bile salt-sterol-glyceride emulsion to gel columns yielded an excluded emulsion phase and a partially included micellar phase. A two-phase oil-micellar system at TG and FA is 6 mM sodium taurodeoxycholate. With increased concentration in glyceride-FA emulsion the bile salt-cholesterol mixed micelles had a larger size. Sitosterol

Table XXXII. Partition coefficient (micellar phase/emulsion phase, M/E), intestinal up-take, and intact animal (rat) absorption of sterols. After Borgström [135]

Sterols	M/E	Intestine	Intact animals
Cholesterol	0.25	1.0	50
Cholesteryl methyl ether	0.068	2.5	41
Cholesteryl propyl ether	0.047	2.0	11
Cholesteryl decyl ether	0.021	2.4	7
Sitosterol	0.23	1.1	5

lowered the micellar distribution of cholesterol. Conversely, cholesterol reduced the micellar distribution of sitosterol. In vitro experiments with rat and hamster intestinal slices (rings) disclosed that the tests had no direct bearing on the in vivo situation. The uptake of labeled cholesterol and of sitosterol was 14 and 15 mg/100 mg tissue in vitro. This did not conform to their respective percent distribution into a micellar phase.

Millipore filters with different partition coefficients $K_{M/E}$ (M = micelles; E = emulsion) for the micellar phase were used at different bile salt concentrations at pH 6.3. An 'artificial intestine' was made with a filter of 3,000 Å pore size. The dimensions of the micelles were 5,000–10,000 Å in the diet, only 30–40 Å in the intestinal content, and 5,000 Å in the thoracic lymph chylomicrons, apparently after resynthesis by intestinal cells. $K_{M/E}$ decreased with increasing chain length of the aliphatic alcohol. This paralleled the absorption as measured by their appereance in the thoracic lymph of rats. The whole-animal absorption was 10 times higher for cholesterol than for sitosterol, in spite of nearly equal $K_{M/E}$ values (table XXXII). Since the half-life of intestinal cells is 24–36 h, sitosterol which had been absorbed may have been excreted with these cells.

Quantitative aspects of sterol absorption were explored next. When rats were fed with increasing amounts of cholesterol in a constant amount of triolein, the percent of cholesterol absorption decreased only gradually. At the lowest dose, only about 50% of dietary cholesterol was absorbed; with larger doses, more was absorbed. Solubility in fat represents a limiting factor. Sitosterol in triolein was absorbed at only one tenth of the cholesterol amount. The absorption site was fixed as the proximal part of the small intestine. 4-^{14}C-cholesterol and 22,23-^{3}H-sitosterol were taken up at the sames sites, i.e. the apical half of the villi. However, the absorption of sitosterol was 3–8% per 24 h and that of cholesterol was 30–40%. There was no mutual interference in absorption. Thus, it was concluded that the specificity of sterol absorption must lie in the intestinal mucosa itself. More recently [849] it was demonstrated that the absorption depends on the blood supply. With blood supply, the uptake of rat jejunal loops was 8–30% half an hour after administration of cholesterol. Without blood supply, it was only 1.5–4.5%. Uptake of labeled cholesterol was reduced more than uptake of sitosterol, although it was absorbed only in amounts no greater than 7% of cholesterol absorption. Besides blood supply, hypo- or hyperthermia also played a role in sterol absorption when concentrations of bile salts were high. Gradual reduction of the blood supply produced *in vivo* conditions which more and more resembled those obtained *in vitro.* The flow rates of sterols from the micellar phase through the cell membrane of the mucosal cells were altered.

4. Enhanced Cholesterol Excretion: Intestinal Microflora

Another factor concerning the interplay between sitosterol and cholesterol is based on presumptive alteration of cholesterol excretion. References to the interference of intestinal bacteria with sterol metabolism were made in earlier sections. Instead of repeating these, let us reiterate that the type of microflora involved in modifying sterol metabolism have never been properly identified and that only scant attention has been paid to changes in the flora subsequent to changes in the diet.

Kuksis et al. [514] analyzed 12 fecal samples from 3 adult men during the last 6 days of a 12-day dietary experiment. Inclusion of 14 g/day sitosterol in a fat-free diet increased the excretion of cholesterol from 0.3 to 1.8 g/day. It lowered bile acid excretion from 0.7 to 0.3 g/day, increased free FA excretion from 0.2 to 6.7 g/day and esterified FA from 0.6 to 0.8 g/day, as analyzed by TLC. Addition of 1.2 g/day of sitosterol to a high-butterfat diet increased cholesterol excretion from 0.6 to 0.8 g/day, decreased bile acid excretion from 0.4 to 0.3 g/day, increased free FA from 0.4 to 0.5 g/day and esterified FA from 0.4 to 0.9 g/day. Ingestion of sitosterol plus butterfat increased the conversion rate of cholesterol to coprostanol about 20-fold. Sitosteryl esters accounted for much of the fecal steryl esters. All these changes were ascribed to the activity of the intestinal microflora.

Cohen et al. [179] found different conversion rates of cholesterol to coprostanol in rats fed cholesterol, sitosterol, sodium taurocholate or sodium taurochenodeoxycholate. The differences were ascribed to bacterial degradation of cholesterol.

Table XXXIII. Effect of sitosterol on cholesterol synthesis in rats

System	Sterol	Effect	Ref.
In vivo	1–2% β-sitosterol diet	none	331, 332
In vivo	2–4% β-sitosterol diet	reduced	728, 729, 734
In vivo	soy sterol diet	enhanced	279
In vivo	sitosterol 5 mg/day i.p. in NaCl for 40 days	enhanced	298
In vivo	β-sitosterol	enhanced	909
Liver slices	2% sterols	enhanced	279, 280
Liver slices	2.5% soy sterols	no effect	876
Liver slices	β-sitosterol: campesterol 6:4 after cholesterol diet	no effect feedback	439, 518

5. Interference with Cholesterol Synthesis

The effect of sitosterol on cholesterol synthesis has been examined in a number of laboratories. The results of studies on the influence of sitosterol on cholesterogenesis in rats are summarized in table XXXIII.

In most instances cholesterol synthesis was enhanced. *Gerson et al.* [298] administered sodium [1-^{14}C]-acetate, and measured $^{14}CO_2$ excretion as well as cholesterol distribution in the rat. They observed reduction of cholesterol content of the aorta, adrenals, plasma, liver and muscle and increase in the cholesterol content of the liver and skin. They concluded that the rate of oxidative degradation of cholesterol was greater than that of its biosynthesis. *Tomkins et al.* [876] studied the effects of feeding a variety of steroids on cholesterol synthesis by rat liver slices. Incorporation of ^{14}C-acetate into cholesterol was inhibited by cholestenone, 7-dehydrocholesterol, lathosterol and dehydroisoandrosterone. Ergosterol, pregnenolone, dihydrocholesterol, 7-ketocholesterol, diosgenin and mixed soy sterols were without effect.

The accounts by *Werbin et al.* [916] on the conversion of s.c. injected ^{3}H-β-sitosterol to ^{3}H-cortisol in guinea pigs do not fit the picture as presented by others. Such conversion would require multiple intermediary steps which very probably include cholesterol.

Zilletti et al. [953, 954] found that sitosterol feeding increased hepatic cholesterol synthesis in mice. *Grundy et al.* [345] found that sitosterol administration to human subjects greatly reduced absorption of exogenous and endogenous cholesterol, with a compensatory increase in cholesterol synthesis.

6. Enzymatic Control of Cholesterol Synthesis

Those who subscribe to the concept that sitosterol affects cholesterol synthesis must consider its effects on the enzymes of cholesterogenesis. In their sterol balance studies, *Raicht et al.* [679] fed rats 0.8% β-sitosterol and 1.2% cholesterol. Sitosterol inhibited cholesterol absorption. In the homeostatic process, cholesterol synthesis by rats was enhanced by 20–28.8 mg/day. There was no change in bile acid formation. There was fair correlation between HMG-CoA-reductase with the cholesterol/bile acid balance. HMG-CoA reductase activity was inhibited by 80%, hydrolysis was enhanced by 61%. In rats fed β-sitosterol, HMG-CoA reductase increased but hydrolysis remained the same as in controls. Despite adaptation to cholesterol feeding there was accumulation of cholesterol in the liver, from 2.2 to 9.2 mg/g.

Cohen et al. [179] fed rats a stock diet plus 1% cholesterol or 1% sitosterol. Daily cholesterol synthesis was 18 mg on the basic diet, nil on cholesterol feeding, 25 mg on sitosterol intake. Bile acid synthesis was 11.8, 22.4, and 12.4 mg, respectively. Biliary cholesterol concentrations were 0.21, 0.15, and 0.28 mg/dl. Bile acid synthesis correlated with 7α-hydroxylase activity and doubled on the cholesterol diet.

Shefer et al. [759] pointed out some limitations in drawing conclusions from acetate incorporation. In rats held for 12 h in darkness there was a dark cycle with peaks of acetate incorporation 6–8 h after the start, i.e. between midnight and 02.00. Intraduodenal infusion of Na taurocholate at a rate of 14 mg/dl/h/rat (to stimulate the bile acid pool) reduced the 100-mg HMG-CoA reductase to the activity of controls. After 2 and 3 days, the activity in controls was 95.8 and 105.0, respectively, 102.9 and 107.3 on infusion of bile acids, 209.1 and 331.5 in bile fistula rats and, finally, 111.2 and 118.2 on bile acid infusion to bile-fistula rats. Intestinal HMG-CoA reductase exhibited a diurnal rhythm which paralleled that of the liver but had a considerably lower amplitude. The specific activity of HMG-CoA reductase increased by 60–70% from a diurnal minimum between 09.00 and noon to a nocturnal maximum between midnight and 02.00. This persisted in cholesterol-fed rats, while in rats receiving bile acids the nocturnal peak was abolished. On combined cholesterol-bile acid feeding, HMG-CoA reductase decreased to 30–40% of the basal level. Since bile acid excretion was nearly the same in rats fed cholesterol and bile acids and in rats fed only bile acids it was concluded that inhibition of HMB-CoA reductase was not the result of intestinal bile acid flux. It could have been due to increased concentration of cholesterol in crypt cells. Rats fed 2%

sitosterol had significantly increased intestinal HMG-CoA reductase, but not when they received sitosterol together with Na taurocholate or Na taurochenodeoxycholate.

Weigand [906] stated that sitosterol and cholesterol compete for absorption, that sitosterol mobilizes cholesterol and that it disturbs homeostasis. The speed of cholesterol synthesis is determined by the conversion of HMG to mevalonic acid. This change is controlled by HMG-CoA reductase. A combination of cholesterol-mobilizing sitosterol with a drug which blocks cholesterol synthesis would be most effective in controlling cholesterol levels. He supported his claim by citing results of experiments with rats. Rats were fed laboratory ration and corn oil (2% plant sterols) at about 400 mg/day, together with 1% bile salts at 200 mg/day. After 1 week, bile was collected by fistula for 30–60 min and liver microsomes were analyzed. For rats fed only 2% β-sitosterol and for those fed sitosterol plus cholic acids the respective results were: liver cholesterol, 2.25 and 2.30; bile cholesterol, 0.51 and 0.20; cholesterol-7α-hydrolase, 11.6 and 1.8; HMG-CoA reductase, 0.309 and 0.094; and bile flow, 6.86 and 12.69.

Kritchevsky et al. [504] studied the effect of a number of hypocholesterolemic drugs on the ratio of activities of aortic cholesteryl ester synthetase to cholesteryl ester hydrolase. Seven groups of rabbits were used. One group was fed a basic ration, the second was fed the ration augmented with 5% corn oil and in a third group the ration was augmented with 5% corn oil and 1% cholesterol. Four other groups were fed the cholesterol-corn oil diet plus one of the following drugs: nicotinic acid (0.5%); *D*-thyroxine (0.5 mg/day); ethyl-β-chlorophenoxyisobutyrate (0.3%), and β-sitosterol (1%). Addition of corn oil to the basic ration increased the synthetase: hydrolase ratio by 64% and when cholesterol was present the synthetase:hydrolase ratio was increased by 177%. All the agents gave synthetase:hydrolase ratios lower than those observed in the aortas of the cholesterol-fed rabbits. The synthetase:hydrolase ratio observed in the rabbits fed β-sitosterol was the same as that seen in rabbits fed only corn oil.

7. Tissue Culture Studies of Cholesterol Synthesis

Avigan et al. [42] used 20 cell lines of human skin fibroblasts obtained from punch biopsies from persons who were free of lipid metabolic disorders. The cells were preincubated for 18 h with lipid-free serum, washed, and incubated with labeled sterol precursors for 4 h at 37 °C. Non-saponifiables and FA fractions were extracted for radioactive assay. Incorporation of ^3H_2O reflected the rate of enzymatic synthesis from various precursors. The presence of either β-sitosterol or cholesterol in the nutrient medium reduced the biosynthesis of cholesterol.

Mouse L cell fibroblasts were used as test cells by *Rothblat and Buchko* [708] and *Rothblat and Burns* [709]. Cholesterol, desmosterol, lathosterol, 7-dehydrocholesterol, and cholestanone all reduced *de novo* synthesis of cholesterol at high concentrations of exogenous cholesterol. At levels of 5, 20, and 40 µg/ml, none of the following: β-sitosterol, stigmasterol, campesterol, ergosterol, cholesteryl oleate, and cholestane, had a marked effect on cell growth or cholesterol biosynthesis. Coprostanol and $Δ^4$-cholestenone were toxic to cells. Mouse L cell fibroblasts grown in absence of exogenous cholesterol were found to synthesize sterols, especially desmosterol. On addition of cholesterol to the nutrient, cholesterol was incorporated which elicited a feedback response that blocked acetate incorporation. While coprostanol and $Δ^4$-cholestenone inhibited sterol synthesis and cell growth, C_{28} and C_{29} phytosterols (sito-, stigma-, campe- and ergosterol) did not. Cells incorporated less than 3% of either cholesterol or β-sitosterol, although this was sufficient to depress synthesis. The higher the levels of sterol in the medium the greater their incorporation into cells. After 24 and 48 h incubation, the results of sterol incorporation were 6.8 and 8.5% for cholesterol, 2.3 and 3.0% for sitosterol, and 4.2 and 5.8% for cholesterol plus sitosterol. Sitosterol was incorporated to one third of the level of cholesterol incorporation into cells. A maximum sterol level of 40 µg/dl in the culture medium reduced sterol synthesis by 80%. Of particular interest was the incorporation of sterols into subcellular fractions (table XXXIV).

Pollak [666], commenting on tissue culture studies, found them of great interest, especially the much lowered incorporation of sitosterol into the cells and the lowered incorporation on addition of cholesterol plus sitosterol to the nutrient. These studies should be used as a basis for further experiments using various S:C ratios. They also should be repeated with human jejunal cells which, *in vivo,* are the site of sterol absorption.

8. Competitive Esterification

The question as to whether or not cholesterol is or can be absorbed from the intestinal lumen of mammals only in its free form or only after esterification has not yet been settled. The wording of the above question may be wrong if we agree that complex micelles, of which cholesterol is a

Table XXXIV. Average incorporation of sterols into subcellular fractions of mouse L-cell fibroblasts (µg/ml/48 h). After *Rothblat and Burns* [709]

Substrate	Cholesterol Cholestanol		Cholesterol Sitosterol	
Culture medium	1.00		1.03	
Whole cell	1.34		1.75	
Plasma membrane	1.30		2.14	
Mitochondria	1.31	1.33	2.05	2.10
Microsomes	1.39		2.10	
Nuclear fractions 1	1.33		1.35	
Nuclear fractions 2	1.39	1.37	1.51	1.32
Nuclear fractions 3	1.39		1.20	
Supernatant 100,000 g	1.49		1.66	

part, are being absorbed. The question would then concern the participation of free cholesterol (FC) and/or of cholesteryl esters in the micelle formation.

There is, however, agreement on the role of enzymatic regulation of both the hydrolysis of steryl esters and the esterification of free sterols. The enzymes originate in the pancreas and in the intestinal wall. They act within the intestinal lumen, most probably on the surface of jejunal mucosal cells.

Hernandez and Chaikoff [379, 380] and *Hernandez et al.* [381] stated that the capacity for hydrolysis and synthesis resides in a single enzyme. They felt that free SH groups of glutathione and cysteine are involved in the enzymatic esterification of cholesterol; further, that β-configuration of OH at C_3 is essential for esterification. They supported this concept by experiments. Rats were fed 4-^{14}C-cholesterol or epicholesterol. In thoracic fistula lymph, 50% FC and 50% EC was recovered, but only free epicholesterol. 20 rats were given 4-^{14}C-cholesterol in cottonseed oil ± 95% phytosterols (10–15% stigmasterol and 75–80% β-sitosterol) or 25 mg β-sitosterol. Thoracic duct lymph was collected for 48 h. About 30% of the ^{14}C appeared in the lymph when cholesterol was fed alone, but only 5% when cholesterol was fed with phytosterols. The fact that less of the recovered labeled cholesterol was esterified in the lymph of rats fed soy sterols or sitosterol was interpreted as evidence of interference of these sterols with cholesterol esterification.

The term 'interference' was changed to 'competition' by *Swell et al.* [834]. They added soybean sterols at several levels to a semi-synthetic diet of rats. There was a dose-related response: at 21 days, the average serum cholesterol levels of rats whose diets contained 0.5, 1, 2, 3, or 4% sitosterol were 90, 79, 59, 50, and 44% of the control values. *In vitro* experiments showed that pancreatic esterase esterified sitosterol, although much more slowly than cholesterol, with either oleic or butyric acid [837]. It was concluded that: (1) the esterifying activity is influenced by the number of double bonds and by the structure of the side chain (e.g. β-sitosterol without a double bond in the side chain-enhanced esterification more than stigmasterol), and (2) the hydrolytic activity is diminished by the ethyl group in the side chain (e.g. cholesterol is hydrolyzed to a greater extent than β-sitosterol). *Swell et al.* [834, 836, 846, 847] contributed many essays to the topic of enzyme involvement in sterol metabolism. They found that in the rat a metabolic pool of FC exists in the intestinal mucosa prior to esterification and its transfer to lymph. Experiments with pancreatecto-mized rats demonstrated that bile salts are required for enzyme activity. *Swell and Treadwell* [842–844] postulated that the function of pancreatic esterase depends on: (1) the temperature at time of exposure; (2) the length of exposure; (3) the concentration of the enzyme (extract); (4) the amount of substrate (cholesterol), and (5) on the pH. Enzymatic function decreased with increased temperature, length of exposure, concentration of the extract, and amount of cholesterol, although here the decrease was less. The optimal pH varied; it was 6.1 for oleate, 4.7 for acetate. *Vahouny and Treadwell* [885] concluded that dietary cholesteryl esters are hydrolyzed in the intestinal lumen prior to the absorption of the moiety as FC. *Subbiah* [802] also stated that complete hydrolysis of cholesteryl esters precedes absorption, in a discussion of the partition coefficient between micellar and oil phase.

Murthy and Ganguly [603] fixed the optimum pH for hydrolytic activity for cholesteryl oleate as 9.6, the optimum pH for synthetic activity of rat pancreas and intestine as 6.1. The two functions were independent of each other. They had different sensitivity to additives: deoxycholate in-hibited the synthetic activity but not the hydrolytic activity. The esterif-ication of β-sitosterol by cholesterol esterase was 86% under the influence of the intestinal enzyme but only 35% with the pancreatic enzyme. Stig-masterol was esterified 5% with intestinal and 4% with pancreatic enzyme. With emphasis on pancreatic cholesterol esterase as the catalyst of rapid synthesis of β-sitosteryl oleate and cholesteryl oleate, *Korzenowsky et al.* [491] tested β-sitosterol with cholesterol. The percent of 4-^{14}C-cholesterol which was esterified decreased with increasing amounts of β-sitosterol.

Schön and Engelhardt [732] fed rats for 1 week on a diet with 2 or 4% sitosterol as a 20% emulsion (containing a small amount of β-sitosterol) plus 25% oleic acid. The 'apparent' absorption of the 2% additive was 56% and that of the 4% additive was 78% as judged from the result of the color reaction of fecal sterols. Phytosterols were thought to inhibit cholesterol absorption by blocking cholesterol esterase.

To add to the controversy, *Shirotari and Goodman* [765] reported that 65–70% of the absorbed ^{3}H-label of cholesterol was found in cholesteryl palmitate, linoleate and oleate in feces, though at different levels becaue of different degrees of hydrolysis. They maintained that hydrolysis is a prerequisite for cholesterol absorption.

Kuksis and Huang [516] fed various phytosterols to dogs. By GLC of the thoracic lymph collected for 18–25 h they observed maximum phytosterol absorption, i.e. 20% of total lymph sterols, at the same time that maximum cholesterol absorption occurred, namely in 8–12 h. In contrast to cholesterol, absorbed phytosterols were present chiefly in the free form. This was interpreted as evidence against the esterification concept.

Daskalakis and Chaikoff [212] also opposed the idea that plant sterols interfere with cholesterol esterification. They fed rats ^{14}C-cholesterol with or without other sterols and collected thoracic lymph for the first 24 h. Allocholesterol, cholestenone, dihydroandrosterone, 7-dehydrocholesterol, epicholesterol, brassicasterol, and stigmasterol were totally ineffective in decreasing absorption. Dihydrocholesterol and ergosterol greatly reduced the amount of absorbed cholesterol – without altering the usual (i.e. 70%) amount of esterified lymph cholesterol.

The question of sterol specificity of hydrolase – which, if answered in the affirmative, would discredit the whole concept of competitive esterification – was studied by *Vahouny and Treadwell* [886]. In their 1968 review of the subject they postulated that cholestanol is as readily esterified with oleic acid as is cholesterol. This process of esterification is slower for β-sitosterol or β-sitostanol. Ergosterol, 3α-hydroxy-cholest-5-ene, 3β-hydroxy-cholest-5,7-diene, 3α- and 3β-hydroxy-5β-cholestane are all very poorly esterified.

Kothari and Kritchevsky [492] studied esterification of oleic acid with various sterols using aortic cholesteryl ester synthetase from rats and rabbits. Cholesterol was esterified to roughly the same extent as cholestanol. The esterification of β-sitosterol was about 20% that of cholesterol.

The evidence seems to weigh in favor of the concept that hydrolysis (not esterification) is a prerequisite for the absorption of cholesterol.

9. Acceptor Sites

Closely linked with the topic on esterification of sterols is the question concerning receptor sites on or within the intestinal mucosal cell membrane.

Glover et al. [304–307, 309, 310] contributed several reports on this subject. They used guinea pig intestinal mucosal cells for the study of absorption of cholesterol and 7-dehydrocholesterol in 150-mg doses. According to them, absorption is a process which – on the molecular level – is the transfer between cell membrane LPs, organelles, and cytoplasm. Stereochemistry is the clue to the poor absorption of phytosterols. They do not fit on the acceptor LPs as readily as cholesterol. They partially reduce cholesterol absorption. Since no esterified cholesterol or 7-dehydrocholesterol was found in the cell fractions, mitochondria, microsomes, nuclei, and debris, the authors concluded that esterification is irrelevant for the mechanism of acceptor site attachment. A year later, however, they stated that esterification is a prerequisite for cholesterol absorption. They agreed that hydrolysis of neutral fat and cholesterol takes place prior to absorption. Bile acids increase absorption – cholate more than deoxycholate. Fats enhance absorption, as FAs, the latter more than neutral fat (NF). The cholesterol molecule attaches itself to animal LPs quite rapidly, whereas phytosterols do so much slower at first and also progressively slower if they do so at all. When a phytosterol molecule succeeds in attaching itself to animal protein the transfer center becomes temporarily immobilized. Liberated excess sterol now becomes esterified and a smaller amount of free sterol than of esterified sterol enters the intestinal lymph. Overload of the system leads to esterification within the mucosa. In another study, guinea pigs received recrystallized cholesterol and 7-dehydrocholesterol while on a fat-free diet for 3 days. After 1 day of fasting, they were given 150 mg/ml arachis oil and 5 g of the fat-free diet for another day and finally on the last day, 150 mg/ml arachis oil plus sterols. Normally, there are no steryl esters in the intestinal mucosa of the guinea pig, but on feeding cholesterol, esters were occasionally found. The authors arrive at a reconciliation of conflicting statements in the following way: esterification is not involved in sterol absorption across the outer cell membrane, and esterase is not obligatory for absorption although it accelerates the process. Similar conclusions were reached by *Clayton et al.* [176] in regard to sterol absorption by roaches.

Ashworth and Green [34] studied the uptake of chelate-dispersed lipids by human αLP. These LPs contained more esterified than free sterols. They are capable of the uptake of 15 times the amount of labeled 7-dehydrocholesterol than of the corresponding ester. *In vitro,* 7.5 mg of unlabeled ergosterol, β-sitosterol, or stigmasterol were mixed with 2.5 mg ^{14}C-cholesterol. This resulted in a lower uptake of cholesterol.

LPs may be able to bind cholesterol *in vitro*. It is by no means certain that they are the acceptor sites of sterols on the cell membrane. The role of membrane phospholipids and related compounds seems more important.

10. Membrane Permeability: Kinetics of Sterols

The interference by sitosterol with cholesterol metabolism has been explained in various ways. Whichever concept one accepts, the fact remains that cholesterol has to traverse the membrane of intestinal mucosal cells. Thus, we are concerned with the events which take place on and within cell membranes and with transmembrane kinetics.

It seems unavoidable to refer here to some essays and texts on biologic membranes, although they do not always refer to phytosterols. Much of the data derive from observations on whole plants and *in vitro* studies with plant cells, and some extrapolation from plant to animal cells may be feasible.

Kúthy [523] demonstrated that phytosterols are as diffusible through parchment membranes as cholesterol. However, in contrast to cholesterol, large amounts of plant sterols in the human diet do not increase the amount of sitosterol in the organs.

Rouser et al. [710] discussed the differences in the structure of cell membranes. Stabilizing of lipid bilayers is accomplished by tail-to-tail interaction of molecules. SFAs lie together adjacently, as do UFAs and chains with the same number of double bonds. *Chapman* [163] and *Chapman et al.* [164] provided additional information. Mixed monolayers of cell membranes are the arena for interaction between cholesterol and phospholipids. The action of cholesterol is related to the physical state of pure phospholipid monolayers. The effect of cholesterol on gel-liquid crystal transition of lipids can be monitored by the caloric method, by NMR, electron spin resonance (ESR), IR, and circular dichromism spectrum (CDS). In the presence of two parts of cholesterol and one part of lecithin, the first eight carbon atoms of the bilayer are rigid, the remaining carbons greatly increasing their motion towards the center of the bilayer. The role of cholesterol lies in the formation of an intermediate, fluid state with several types of lipid classes. The amount of cholesterol in cell membranes differs. This is most important with regard to permeability. One must further differentiate between 'liquid' and 'solid' membranes.

Cole [181], reviewing the IR spectra of natural products, stretching absorption, and the stereochemistry of steroids, refused to use IR alone to measure special arrangements of the molecular skeleton and substitution groups in asymmetrical large molecules. Bending vibrations are excursions out of the plane. They are greatly influenced by substitutes of the ethylene system. Stretching vibrations of methyl and methylene groups cause strong spectral absorption. There are differences between C-C, C=C-H, and C-H stretching vibrations of cis-di-substituted double bonds in steroids, each with a characteristic frequency. Stretching vibrations of C-OH bonds generally depend on the stereochemical structure: the frequency is higher for axial than for equatorial compounds. The axial groups are thermodynamically more stable than the equatorial groups. During vibration, C_3 moves toward the center of the ring in the equatorial type. The resulting widening of the ring angle produces a greater restoring force – higher frequency – compared to the axial compound, when the rotation involves a perpendicular bending of the ring which can be distributed to the A/B junction. The L-shape of the A/B cis-steroids makes the ring system more rigid and it is responsible for smaller difference between the equatorial and the axial frequencies compared to A/B trans-sterols.

Sallee and Dietschy [718] discussed the uptake across the jejunal brush border and the relative roles of the unstirred water layer and the lipid cell membrane as determinants of the absorptive process. This diffusion barrier in the intestine can be considered as an unstirred water layer. The authors stated that the uptake sites for FA are linear with respect to concentration of the fat solute, that there is no competitive metabolic inhibition or contralateral stimulation, and low temperature dependence. Uptake proceeds by passive diffusion. The apparent permeability coefficient is depressed by bile acid micelles in proportion to FA chain length. The uptake of short-chain FA monomers is rate-limited by the lipid cell membrane, but the diffusion through the unstirred water layer becomes increasingly rate-limited with increase in FA chain length. This concept offers a plausible explanation for the role of PUFA and β-sitosterol, or of linoleic acid and β-sitosterol, as two sets of hypocholesterolemic factors in vegetable oils.

Other investigators have paid more specific attention to phytosterols. *Atallah and Nicholas* [35] incubated cell-free extracts of *Phaseola vulgaris,* beans, with 2-^{3}H-cholesteryl palmitate, 22,23-^{3}H-sitosteryl palmitate, cholesterol, and sitosterol. Three types of structures evolved: (1) turbid, viscid, soap-like smectic structures of several types, demonstrated by XD; (2) normal, turbid, mobile mesophase – liquid crystalline state nematic structures, or (3) twisted, turbid, mobile, threaded, abnormal cholesteric structures. The side-chain structure and the positive double bonds in the nucleus were considered critical in determining whether or not a steryl ester came from a liquid crystal state, a mesophase. Mesophilic formation of thermotropic liquid crystals is influenced by heat, electric current, pressure, and impurities (especially those from solvents). The dependence on temperature was demonstrated by different degrees of incorporation of cholesterol and sitosterol and their palmitates into mitochondria and microsomes at different temperatures.

Let us recall the reports by *Smith* [776, 777] from an earlier section. He postulated the requirement for a planar molecule with the 3-OH group in the equatorial position for sterol penetration of the cell membrane of mycoplasma. *Grunwald* [348, 349] experimented with red beet tissue and barley roots and came to the conclusion that permeability required a free OH-group on C_3 and that the bulky molecules of stigmasterol and β-sitosterol were not capable of penetrating the cell membrane. Campesterol penetrates better than the other plant sterols, but not as well as cholesterol.

Studies with bacteria and cells from higher plants provided a background for studies with animal cells. *Edwards and Green* [246] worked with intestinal cells. They prepared (Goad's) lysosomes which were composed of bilayers of egg lecithin, and added to these 70.7% sitosterol and 29.3% campesterol. They also prepared lysosomes with 58% phytosterols, 27% lecithin and 15% cholesterol. Incorporation of phytosterols into erythrocyte lysosomes was slower than incorporation of cholesterol. Phytosterols are poorly absorbed, being unable to enter the cell membrane and the soluble LPs of intestinal cells. At this point they may block the entry of cholesterol.

When *Glover and Green* [305] examined intestinal cells of guinea pigs, they found 20% of sterols in the cell membrane, the nuclei and mitochondria, 40% in microsomes, and less than 0.5% in cytosol and fat globules. This is reminiscent of sterol distribution in plant cells, following passage through the cell membrane. *Glover and Stainer* [309] recovered from male rats fed on a fat-free diet, 15% of total steroids, and from fasting rats, 10% of total steroids in the mitochondria of mucosal cells. About half of the total steroids were in the mitochondria and in the microsomes, 30% in the cytosol, 10% in the nuclear fractions and in debris.

Favarger and Roth [270] ascribed the low resorbability of vegetable sterols and saturated sterols to their inability to penetrate the external membrane of intestinal epithelial cells. Addition of ergosterol to a sterol-free diet of rats – a diet containing 0.15–0.2% cholesterol in olive oil – lowered the coefficient of digestibility of the olive oil by 10%. Addition of sitosterol had no effect. The digestibility of olive oil for rats fed on a diet containing 0.25–2% cholesterol was 32–48%. This was lowered by ergosterol. Three explanations for this were offered. (1) The majority of droplets which compose the intraluminal emulsion undergo an even greater dispersion in the mucoid layer. Mucoproteins and phospholipids form LP complexes which probably exchange their lipid components with that of the cell membrane LP. (2) Cholesterol esterified in the organism probably participates in the hydrotropic system. (3) Vegetable sterols and saturated sterols find no interactive response in the cell membrane and are repelled by the membrane system. *Favarger* [269], reviewing the absorption of sterols in 1958, reemphasized earlier observations and conclusions. Referring to β-sitosterol, he stated that it is difficult to fix its coefficient of digestibility. He ascribes its low digestibility to side chain modifications or to interversion of H and OH at C_3.

To summarize is to emphasize the stereochemical interplay between various sterols and cell membrane constituents, mainly phospholipids and cholesterol. In simple terms, the 'bulkiness' of phytosterols seems to be the obstacle to their penetration of the cell membrane, and thus to absorption. The adhesion of phytosterols to the cell surface may block the penetration by cholesterol. Conjugation of β-sitosterol with cholesterol, whether as clathrates or as mixed crystals or complex micelles with bile salts and FAs of varying chain length, deserves further consideration.

Part IV

A. Summary of Clinical Studies with Sitosterol

This section was deferred to allow the inclusion of data interwoven in reviews of various types of hypercholesterolemia, factors influencing the results of studies, and data concerning sterol absorption.

The difficulties encountered in the analysis of the bibliography apply also to this section. Many references to clinical trials with sitosterol have been made in essays on the use of vegetable oils. Thus, one cannot determine the exact number of reports concerning the application of phytosterols and/or sitosterol. Since many reports lack detailed information, it is also impossible to determine the number of persons for whom sitosterol has been prescribed. Moreover, multiple reports have been issued on the same group of subjects, or new patients were added to old ones, leading to distortion of total numbers of patients. By careful scrutiny of all data, one reaches a total of approximately 1,800 subjects referred to in the literature on clinical use of sitosterol as a cholesterol depressant. Surely, the number must be far higher since not all the physicians who used sitosterol reported on the results. Of the 1,800 subjects, 47% were apparently 'healthy' and 'normocholesterolemic' or had only mild elevation of plasma cholesterol levels, and 53% were referred to as 'patients' with hypercholesterolemia and often with clinical evidence of atherosclerosis. Some reports concerned a single patient. The largest number of patients in a report was 118. The average number of subjects was 20 per report.

It is difficult to depress a 'normal' plasma cholesterol level and largely unsuccessful. It is also unnecessary. Actually, the overall results of clinical trials are surprisingly good if one realizes that nearly one half of all the subjects (47%) of an unselected pool had normal cholesterol levels. Most of the research teams reported complete success, i.e. response of all their subjects, in spite of a wide range of plasma cholesterol values. Only a few investigators met with complete failure – for reasons discussed elsewhere. Positive effects of sitosterol ranged from 50% of the probands (in most studies) to 100%. The figure of 50% is close to the percentage of patients with hypercholesterolemia (53%).

Because of the lack of selection of subjects, the results varied with regard to those who responded and the degree of their response. A statistically significant reduction – usually by 10–12% or over, of the plasma cholesterol from the pretreatment level – was reported for about 66% of all the subjects. 'All the subjects' include those with normal plasma cholesterol levels, those with moderate elevation, those with fluctuating levels, and also those with monogenic hypercholesterolemia. Again, considering this mixed population, the results with regard to the degree of the effect were surprisingly good. For all the subjects, the average plasma cholesterol decrease was 18%. Maximum decreases were reported as 40% for 'healthy' subjects and 60% for 'patients', i.e. in 1 patient. Averages have limited significance because of the wide span (1–60%) of the fall and because the average depends on the distribution of values within the range of each series.

All the investigators who checked the plasma cholesterol level before the start of the sitosterol regimen – and not all did so, or did not report on it – agreed to two features: (1) the effect or the lack of it becomes apparent within 2 weeks of the regimen and (2) the effect depends on the height of the pretreatment plasma cholesterol level.

The fact that the effect becomes apparent after a short course of treatment allows determination as to whether the regimen should continue and also determination of the proper dosage of sitosterol. After a second 2-week course, the daily dose can be increased or decreased to a maintenance dose. Biweekly monitoring of the plasma cholesterol level also reveals stability or instability of the patient's cholesterol.

The results of 'sitosterol therapy' are fairly predictable, assuming that a selection of patients has been made. Monogenic hypercholesterolemia, especially if of the homozygous type, will not respond, or will respond only poorly and/or only temporarily. In polygenic hypercholesterolemia, one will treat the underlying cause to bring about a reduction in plasma cholesterol, in spite of the observation that patients with hypothyroid states and those with nephrosis respond fairly well to sitosterol ingestion. This leaves us with those in whom hypercholesterolemia is of dietary origin as the group most likely to benefit from the sitosterol regimen. It is the largest group of subjects with hypercholesterolemia in the Western World and the group includes most patients with atherosclerosis.

The statement that those with higher plasma cholesterol levels respond to sitosterol ingestion with a greater reduction than those who have only a moderately elevated level has its limitations. Patients with monogenic hypercholesterolemia and those with hypothyroidism have the highest plasma levels, and the former are exceptions to the rule, i.e. of reciprocity between pre- and posttreatment cholesterol levels.

Sitosterol is not a drug but a natural constituent of many foods. It becomes a pharmaceutical if it is dispensed as such. It has few, if any, side effects and no known adverse ones. It does not become ineffective in the course of prolonged intake over several years. It can easily be combined with other dietary regimens.

Except for those patients, including some with alimentary hypercholesterolemia, whose plasma cholesterol is marked by lability, the results of sitosterol medication will be satisfactory. Among those with fluctuating cholesterol are patients with diabetes mellitus and patients who had recently had a myocardial infarct. A 20% reduction of a 300-mg/dl level brings the cholesterol to the upper normal limit (240 mg/dl). The majority of individuals who have diet-induced elevation of plasma cholesterol have levels in the 240- to 300-mg/dl range. Cholesterol depression can be achieved with a relatively small daily dose (3–3.5 g/day) of a high-purity preparation (93% β-sitosterol) taken during meals. Such depression will be statistically and, possibly, biologically significant. It can be sustained as long as the regimen is maintained. Of course, it has to be maintained 'forever', as long as the patient continues to eat.

B. The Demise of the First Sitosterol Era

On reading the previous section, the summary of the results of clinical studies using sitosterol as plasma cholesterol depressant, one wonders why this approach had been underplayed. Although there were only a few adverse reports concerning this approach they came from prestigious investigators and thus were accepted uncritically. And they inhibited further investigation.

The main objection to sitosterol, voiced repeatedly in many reviews, was to the 'large amounts of sitosterol required to achieve results'. Thus, one of the latest reviews, by *Abrams and Schwartz* [1] reads: 'Unfortunately, very large amounts must be used and prolonged treatment is expensive.' This and similar statements are based on lack of information. These authors cited only three references, although by 1967 when they wrote their review reports on clinical use of sitosterol exceeded 100. Actually, the amounts of sitosterol preparations used in the first 'sitosterol era', between 1952 and 1964, ranged from 0.3 to 53 g/day, with doses between 1 and 10 g/day used in over half of the studies. In one fourth of all the trials, the dose was less than 7 g/day and the average daily dose used was 4 g/day. These amounts cannot be called excessive, especially if one compares them with the amounts of some other medications. Cholestyramine, a popular drug, is prescribed in 16-gram daily doses. Some of those who objected to the large doses of sitosterol readily gave their patients daily doses of 90, 100, or 120 g of safflower oil, or 100 g of sunflower seed oil, or three daily doses of 350 g each of corn oil (a total of 1,050 g/day). It must be added that the sitosterol preparations used in the past were not of highest purity.

The reference to the high cost of sitosterol medication is hardly worth a comment by anyone who is familiar with the costs and mode of payment for medical care and medications prevalent in the USA.

A major objection to sitosterol ingestion was based on its lack of palatability. As late as 1976, *Borgström* [138] found it proper to quote verbatim a paragraph from *Pollak*'s [663] first report, describing the physical properties and taste of crude sitosterol powder which was taken by the first group of patients. In time, sitosterol preparations, whether dispensed as liquid emulsions, in capsules, as pills, crystals, granules, or as a powder, have become as acceptable as any other drug or dietary supplement.

The need for prolonged treatment has often been mentioned as a disadvantage of the sitosterol regimen. This objection is based on ignorance of the principle of this approach. Multiple dietary measures and pharmaceuticals are being prescribed because they inhibit the absorption of cholesterol from the intestinal lumen. This cholesterol is partly of exogenous and partly of endogenous origin. The exogenous component plays a major role in the alimentary type of polygenic hypercholesterolemia. It plays a minor role in other types of hypercholesterolemia, though everybody ingests some cholesterol. Whether one uses a low-fat/low-cholesterol diet, a vegetarian diet, a diet supplemented with vegetable oils or with phytosterols or pure β-sitosterol, the regimen must be continued for the rest of the patient's life. The same is true for drugs which act on the principle of interference with absorption. On termination of any of these treatments the plasma cholesterol reverts to pretreatment levels within approximately 4 weeks.

The large doses of sitosterol required was also objected to by *Nutrition Reviews* [619], together with the criticism that the effect is modest, a 5–25% drop in the plasma cholesterol level. The quoted range is incorrect and

does not match the analysis of all the reported results. To expect a reduction to a low normal or subnormal level is unrealistic and is biologically not feasible. A 20% reduction brings levels below 300 mg/dl into
the 'normal' range and a 25% reduction does so for any level below
330 mg/dl.

The obvious drawback of sitosterol – again, shared by all measures
which operate on the same principle – lies in the fact that it is not effective
in all patients but only in those with alimentary hypercholesterolemia.
Curiously, this aspect has hardly ever been mentioned.

These adverse comments, though largely unjustified, were not the sole
or even the decisive factor in the demise of the first 'sitosterol era'. The
replacement of 'animal fat' by 'vegetable fat' and the widespread opinion
that the hypocholesterolemic effect of vegetable oils is due to their high
PUFA content became a decisive factor. The National Institutes of Health
and the American Heart Association included in their recommendations
for the prevention of atherosclerosis and its main complications the substitution of vegetable fats for animal fats. They did this without alerting the
medical profession and the public to the fact that not all vegetable oils have
a high iodine number (e.g. coconut oil). They also failed to point out that
the potency of SFA in increasing plasma cholesterol is double the potency
of USFA in lowering the level. Furthermore, it was not stressed that
elimination of saturated fat may be more important than addition of
unsaturated fat to the diet.

Linked to the above-mentioned recommendation of PUFA as the
chosen means of reducing plasma cholesterol was the skillfull advertising
of manufacturers of margarines and shortenings. The American housewife
was and still is being told to replace butter with vegetable oil margarine.
The need for a P:S ratio of 2:1 or, at least 1.5:1, was never mentioned nor
was the fact that the commercial products are made with 'partially hydrogenated' oils. The degree of hydrogenation is not known. It would also be
important to point out that unsaturated fat is a readily identifiable material
whereas sitosterol is not. Thus, recommendations of use of unsaturated
fats can be much more easily followed.

The last factor which led to discontinuation of the clinical use of
sitosterol came with the introduction of many pharmaceutical agents as
cholesterol depressants. Some of these drugs have found wide acceptance.
Much of the public, the American public in particular, seems to prefer
pharmaceuticals to non-medicinal approaches such as a sitosterol regimen
or any other dietary regimen.

C. The Dawn of a Second Sitosterol Era

In 1963, *Hoff and De Graaf* [394] wrote about sitosterol stating that 'there is little interest in this substance at present'. This was true at that time although their implication of finality proved wrong. In the past 4 years several reports on new clinical trials with sitosterol were made from Germany, Canada, and the United States.

The studies by *Lees and Lees* [529] are of particular interest since they compared the effect of emulsions containing 60–65% β-sitosterol with the effect of spray-dried powder of 93% purity. Using various amounts they managed to reduce the daily dose of sitosterol powder to 5 g. Such a dose reduced the plasma cholesterol of 7 patients with type II hyperlipoproteinemia by 16%. Moreover, *Oster et al.* [638, 639] found that when plasma cholesterol of 15 patients decreased by 12.5% ($p < 0.01$), their LDL cholesterol decreased by 19.5% ($p < 0.05$). This is important, since LDL cholesterol has been singled out as 'the' atherogenic component of lipoproteins.

Recently (1976), *Ahrens* [8] summarized the report of several cooperative studies of large numbers (1,103) of patients, with particular reference to the 'Coronary Drug Project'. It appears that the effect of β-sitosterol matches the effect of other measures proposed to reduce plasma cholesterol. In 66% of unselected subjects, β-sitosterol ingestion reduced the plasma cholesterol level by 18%. In those whose level was above 250 mg/dl, the reduction averaged 21%. This compares favorably with the effect of other agents (table XXXV).

The patients who respond to β-sitosterol are the same as those who react to a low-fat/low-cholesterol diet, to a strict vegetarian diet, to a vegetable oil diet, or to cholestyramine. The degree of response to the various measures is also similar. Thus, the physician faces a choice: he can use one agent, combine two, or alternate them. Moderate fat-cholesterol restriction is not very effective and strict restriction is poorly tolerated. Also, compliance of patients with the physicians' instructions is not good. Ingestion of large doses of vegetable oil is not always desirable. Moreover, reports that many persons adjust to such diets and become resistant are disturbing. Although β-sitosterol when taken alone is effective, provided that it is taken in proper form and dose and with meals, a combination of a β-sitosterol regimen with other dietary measures seems to offer a rational approach. In such combination, the diet would not have to be too strict.

Table XXXV. Average reduction (%) of plasma cholesterol by diets and medications

Patients	Diets or drugs	Reduction	Ref.
Unselected	low fat/low cholesterol	−19	*
	strict vegetarian	−13	*
	corn oil	−16	*
	high PUFA	−17	*
TC<212 mg/dl	[high linoleic acid (42%) as corn oil]	−12	840
TC>230 mg/dl		−22	
HLP type IIa	cholestyramine	−20	8
HLP type IIa	cholestyramine and clofibrate	−14	
HLP type IIb	cholestyramine	−17	9
HLP type IIb	cholestyramine and clofibrate	−20	
HLP type II	β-sitosterol	−18	529
TC<240 mg/dl	β-sitosterol	−12	*
TC>250 mg/dl	β-sitosterol	−21	

* Collated data from bibliography.

Hodges et al. [393] compared the customary 'Oriental' and 'Western' diets. The Oriental diet provides adequate protein of mixed vegetable and animal origin, an abundance of complex carbohydrates, relatively little fat, and much less cholesterol than used in the Western diet. On an Oriental diet, the serum cholesterol of 19 healthy Americans became comparable to the lower level of Orientals. The fiber, protein and sterol content of the vegetables may have played some role in their observations. This 'Oriental' diet is in fact a 'Chinese' diet.

Hiscock et al. [391] recommended a palatable diet which was devoid of butter, conventional margarines and hydrogenated shortening, and contained less meat than the usual diet. They preferred corn oil to other vegetable oils. They included in the diet 'filled' milk, composed of 3.5% w/v soybean oil which was homogenized into non-fat milk plus A and D vitamins. They also substituted an 'unsaturated' margarine from natural corn oil, a 'safflower oil ice cream', 'filled cheese', and then allowed seven egg yolks per week. This highly contrived diet was compared to the standard diet (table XXXVI).

The proposed changes amount to a decrease in dietary cholesterol by 50% and a more than fivefold increase in β-sitosterol, a change from a S:C =0.4:1 ratio to S:C=2.3:1. Further, the iodine value was increased twofold

Table XXXVI. Comparison of a cholesterol-lowering diet with a standard diet. After *Hiscock et al.* [391]

Composition	Standard diet	Proposed diet
Total calories/day	2,370	2,430
Protein, g/day	90	94
Fat g/day	107	106
Fat, calories, %	41	40
Iodine number of fat	53	100
Linoleic acid, %	12	38
β-Sitosterol, mg/day	29	160
Cholesterol, mg/day	750	380

and the linoleic acid content threefold. The same result may be achieved by reducing the amount of cholesterol and by incorporating β-sitosterol into the vegetable oil margarine.

This is a book about sitosterol, not about pharmaceuticals. It is, however, impossible to ignore two drugs, Clofibrate or Atromid-S and Cholestyramine or Questran. Clofibrate reduces triglycerides more effectively than cholesterol. The required dose is 2 g/day or four 500-mg capsules. There are, however, increasing numbers of contraindications and adverse effects. Cholestyramine has no effect on triglycerides but reduces plasma cholesterol. The daily dose is 24 g. The list of potential side effects is also long. In contrast, the *Physicians' Desk Reference Book* [656] lists no side effect for sitosterol. In the book, the daily dose, applied to Cytellin, an emulsion with 65% of β-sitosterol, is given as 9 g/day at this writing. The 93% pure powdered product has not yet reached the market. The effectiveness of β-sitosterol and its safety, when compared to other pharmaceuticals, strongly recommends its reintroduction into the clinical practice of medicine.

The dawn of a new 'sitosterol era' is emerging. Sitosterol, alone or combined with dietary measures, offers a non-medicinal approach to the reduction of plasma cholesterol for those in whom its elevation is due to excessive ingestion of cholesterol. This has been aptly summarized by *Lees and Lees* [529]: 'The ideal drug for the treatment of hypercholesterolemia should be effective, free of subjective side effects, and of objective toxicity. The cholesterol analogue, β-sitosterol, comes close to the ideal.'

Since all the measures used to lower plasma cholesterol have a relatively modest effect it is safe to predict that the search for a more 'potent' agent will go on for some time. It is also safe to predict that such attempts will fail. Since it is not feasible to depress plasma cholesterol below the physiologic limit without triggering homeostatic cholesterol biosynthesis it may be prudent to pay at least equal attention to other 'risk fastors' in atherogenesis and thrombogenesis. It is certainly easier to eliminate smoking of cigarettes and the use of oral contraceptives, to control hypertension and body weight than to reduce an elevated plasma cholesterol level to a mean 'normal' range. Efforts to combat other risk factors should not replace but rather supplement attempts to prevent alimentary hypercholesterolemia. To this end, β-sitosterol alimentation presents a rational approach.

References

1   Abrams, W.-B. and Schwartz, M. A.: The pursuit of new antilipemic agents; in Brest, Atherosclerotic vascular disease, pp. 260–273 (Appleton Century Crofts, New York 1967).

2   Abell, L. L. and Kendall, F. E.: Studies on cholesterol metabolism. Effect of diet on sterol excretion. Circulation *4:* 480 (1951).

3   Achor, R. W. P.: Brief survey of the cholesterol problem with clinical application. Minn. Med. *43:* 684–692 (1960).

4   Adami, E.; Marazzi-Umberti, E., and Turba, C.: Anti-ulcer action of some natural and synthetic terpenic compounds. Medna exp. *7:* 171–176 (1962).

5   Agarwal, H. C. and Casida, J. E.: Nature of house fly sterols. Biochem. biophys. Res. Commun. *3:* 508–512 (1960).

6   Ahrens, E. H., Jr.: Nutritional factors and serum lipid levels. Am. J. Med. *23:* 928–952 (1957).

7   Ahrens, E. H., Jr.: A review of the evidence that dependable sterol balance studies require a correction for the losses of neutral sterols that occur during intestinal transit; in Jones, Atherosclerosis, pp. 248–252 (Springer, New York 1970).

8   Ahrens, E. H., Jr.: The management of hyperlipidemia: whether, rather than how. Ann. intern. Med. *85:* 87–93 (1976).

9   Ahrens, E. H., Jr.; Blankenhorn, D. H., and Tsaltas, T. T.: Effect on human serum lipids substituting plant for animal fat in diet. Proc. Soc. exp. Biol. Med. *86:* 872–878 (1954).

10   Ahrens, E. H., Jr.; Hirsch, J.; Insull, W., Jr., and Peterson, M. L.: Effects of dietary fats on serum lipide levels in man, Trans. Ass. Am. Phys. *70:* 224–233 (1957).

11   Ahrens, E. H., Jr.; Hirsch, J.; Insull, W., Jr., and Peterson, M. L.: Dietary fats and human serum lipide level; in Page, Chemistry of lipides as related to atherosclerosis, pp. 222–261 (Thomas, Springfield 1957).

12   Ahrens, E. H., Jr.; Hirsch, J.; Insull, W., Jr.; Tsaltas, T. T.; Blomstrand, R., and Peterson, M. L.: Dietary control of serum lipids in relation to atherosclerosis. J. Am. med. Ass. *164:* 1905–1911 (1957).

13   Ahrens, E. H., Jr.; Hirsch, J.; Insull, W., Jr.; Tsaltas, T. T.; Blomstrand, R., and Peterson, M. L.: The influence of dietary fats on serum-lipid levels in man. Lancet *i:* 943–953 (1957).

14   Ahrens, E. H., Jr.; Insull, W., Jr.; Hirsch, J.; Stoffel, W.; Peterson, M. L.; Farquhar, J. W.; Miller, T., and Thomasson, H. J.: The effect on human serum-lipids of a dietary fat, highly unsaturated, but poor in essential fatty acids. Lancet *i:* 115–119 (1959).

15   Albanese, A. A.; Woodhull, M. L.; Lorenze, E. J., and Orto, L. A.: Dietary fats and blood cholesterol levels in elderly persons. A ten-year study. N.Y. St. J. Med. *65:* 517–529 (1965).

16   Alfin-Slater, R. B.: Factors affecting essential fatty acid utilization; in Garattini and
      Paoletti, Drugs affecting lipid metabolism, pp. 111–118 (Elsevier, Amsterdam 1961).

17   Alfin-Slater, R. B.; Auerbach, S., and Shull, R. L.: Effects of vegetable oils on hepatic
      cholesterol levels. Fed. Proc. 19: 18 (1960).

18   Alfin-Slater, R. B. and Jordan, P.: The effect of safflower oil on the nature of serum
      cholesterol esters. Am. J. clin. Nutr. 8: 325–326 (1960).

19   Alfin-Slater, R. B.; Wells, A. F.; Aftergood, L.; Melnick, D., and Deuel, H. J., Jr.: The
      effect of plant sterols on cholesterol levels in the rat. Circulation Res. 2: 471–475
      (1954).

20   Allais, J. P. et Barbier, M.: Sur les intermédiaires de la déalkylation en C24 du β-
      sitostérol par la criquet Locusta migratoria L. Experientia 27: 506–507 (1971).

21   Allais, J. P.; Pain, J. et Barbier, M.: La déalkylation en C24 du β-sitostérol par l'abeille
      apis mellifica. C. r. hebd. Séanc. Acad. Sci., Paris 272: 877–879 (1971).

22   Altschul, R.: New experiments in arteriosclerosis. Am. Heart J. 36: 480 (1941).

23   Altschul, R.: Selected studies in arteriosclerosis, pp. 150–151 (Thomas, Springfield
      1950).

24   Amsterdam, B.: Fats, cholesterol, lipoproteins, and atherosclerosis; facts, fallacies,
      fancies, and formulas for therapy. N.Y. St. J. Med. 58: 2199–2212 (1958).

25   Anderson, J. T.; Grande, F., and Keys, A.: Safflower oil, hydrogenated safflower oil,
      and ascorbic acid effects on serum cholesterol. Fed. Proc. 16: 380 (1957).

26   Anderson, J. T.; Grande, F., and Keys, A.: Effect on serum cholesterol in man of fatty
      acids produced by hydrogenation of corn oil. Fed. Proc. 20: 96 (1961).

27   Anderson, J. T.; Grande, F., and Keys, A.: Hydrogenated fats in the diet and lipids in
      the serum of man. J. Nutr. 75: 388–394 (1961).

28   Anderson, J. T.; Keys, A., and Grande, F.: The effects of different food fats on serum
      cholesterol concentration in man. J. Nutr. 62: 421–424 (1957).

29   Anderson, R. J. and Moore, G.: A study of the phytosterols of corn oil, cottonseed oil
      and linseed oil. J. Am. chem. Soc. 45: 1944–1953 (1923).

30   Anthony, W. L. and Beher, W. T.: Effects of beta-sitosterol and cholic acid on mobi-
      lization of mouse liver cholesterol. Fed. Proc. 17: 182 (1958).

31   Ardenne, M. V.; Steinfelder, K.; Tümmler, R. und Schreiber, K.: Molekül-Massen-
      spektrographie von Naturstoffen. Steroide. Experientia 19: 178–180 (1963).

32   Armstrong, M. L.; Connor, W. E., and Melville, R. S.: Failure of corn oil and triparanol
      to prevent hypercholesterolemia and atherosclerosis. Proc. Soc. exp. Biol. Med. 113:
      960–963 (1963).

33   Armstrong, W. D.; Van Pilsum, J.; Keys, A.; Grande, F.; Anderson, J. T., and Tobian,
      L.: Alteration of serum cholesterol by dietary fats. Proc. Soc. exp. Biol. Med. 96:
      302–306 (1957).

34   Ashworth, L. A. E. and Green, C.: The uptake of lipids by human lipoproteins. Bio-
      chim. biophys. Acta 70: 68–74 (1968).

35   Atallah, A. M. and Nicholas, H. J.: Function of steryl esters in plants: a hypothesis that
      liquid crystalline properties of some steryl esters may be significant in plant sterol
      metabolism. Lipids 9: 613–622 (1974).

36   Atherinos, E. E.; El-Kholy, J. El-S., and Soliman, G.: Chemical investigation of Cynera
      scolymus. I. Steroids of the receptacles and leaves. J. chem. Soc. pp. 1700–1704
      (1962).

37   Audier, M.; Pastor, J.; Pauli, A.-M.; Poggi, L.; Cloetens, W.; Mascart, P. et De Mey,

D.: Action du béta-sitostérol sur l'athérosclérose et l'hypercholestérolemie humaine. Archs méd. gén. trop. *1:* 3–17 (1961).

38   Audier, M.; Pastor, J.; Pauli, A.-M.; Poggi, L.; Cloetens, W.; Mascart, P. et De Mey, D.: Essai de traitement de l'athérosclérose par le béta-sitostérol. Revue méd. Fr. *43:* 7–8, 11–12 (1962).

39   Auterhoff, H. und Nickoleit, R.: Die Zusammensetzung handelsüblicher β-Sitosterin-Präparate. Acta pharm., Weinheim *305:* 104–108 (1972).

40   Avigan, J. and Steinberg, D.: Effect of corn oil feeding on cholesterol metabolism in the rat. Circulation *16:* 492 (1957).

41   Avigan, J. and Steinberg, D.: Effect of saturated and unsaturated fat on cholesterol metabolism in the rat. Proc. Soc. exp. Biol. Med. *97:* 814–816 (1958).

42   Avigan, J.; Williams, C. D., and Blass, J. F.: Regulation of sterol synthesis in human skin fibroblast culture. Biochim. biophys. Acta *218:* 381–384 (1970).

43   Bae, M. and Mercer, E. I.: The effect of long- and short-day photoperiods on the sterol levels in the leaves of *Solanum andigena*. Phytochemistry *9:* 65–68 (1970).

44   Baisted, D. J.; Capstack, E., Jr., and Nes, W. R.: The biosynthesis of β-amyrin and β-sitosterol in germinating seeds of *Pisum sativum*. Biochemistry, N.Y. *1:* 537–541 (1962).

45   Ballin, J. C.: Answer to question, 'What measures are available for lowering blood cholesterol levels? Are they injurious to health?' J. Am. med. Ass. *175:* 262 (1961).

46   Banerjee, S. and Bandyopadhyay, A.: Further observations on plasma lipids of vegetable oil-fed monkeys. Proc. Soc. exp. Biol. Med. *113:* 541–545 (1963).

47   Banerjee, S.; Ghosh, P. K., and Bandyopadhyay, A. B.: Effect of vegetable oils on plasma lipids of Rhesus monkeys. Proc. Soc. exp. Biol. Med. *109:* 313–317 (1962).

48   Banerjee, S.; Rao, P. N., and Ghosh, S. K.: Biochemical and histochemical changes in aorta of chicks fed vegetable oils and cholesterol. Proc. Soc. exp. Biol. Med. *119:* 1081–1086 (1965).

49   Barber, J. M. and Grant, A. P.: The serum cholesterol and other lipids after administration of sitosterol. Br. Heart J. *17:* 296–298 (1955).

50   Barbier, M.; Hügel, M.-F. et Lederer, E.: Isolement du 24-méthylène-cholestérol à partir du pollen de différentes plantes. Bull. Soc. Chim. biol. *42:* 91–97 (1960).

51   Baron, C. et Boutry, J.-L.: Contribution à l'étude des stérols du plancton méditerranéen. C. r. hebd. Séanc. Acad. Sci., Paris *256:* 4305–4307 (1963).

52   Barr, D. P.; Russ, E. M., and Eder, H. A.: Protein-lipid relationships in human plasma. II. Atherosclerosis and related conditions. Am. J. Med. *11:* 480–493 (1951).

53   Barton, P. G. and Glover, J.: A possible relationship between the specificity of the intestinal absorption of sterols and their capacity to form micelles with phospholipid. Biochem. J. *84:* 53–54P (1962).

54   Bartov, I.; Bornstein, S., and Budowski, S.: The effect of soy sterols in hypercholesterolemic chicks. Poultry Sci. *48:* 1276–1281 (1969).

55   Bartov, I.; Budowski, S., and Bornstein, S.: Anticholesterolemic effect of unsaponifiable fractions of vegetable oils in chicks. 1. Short-term effects of soy sterols. Poultry Sci. *49:* 1492–1500 (1970).

56   Bean, G. A.: Phytosterols; in Paoletti and Kritchevsky, Advances of lipid research, vol. 11, pp. 193–218 (Academic Press, New York 1973).

57   Beeler, D. A.; Rogler, J. C., and Quackenbush, F. W.: Effects of levels of certain dietary

lipids on plasma cholesterol and atherosclerosis in the chick. J. Nutr. *78:* 184–188 (1962).

58 Bergmann, F.; Bandomer, G. und Herget, H. J.: Biliäre Lithogenität und Gallensäure-stoffwechsel unter β-Sitosterin. Z. Gastroent. *14:* 240 (1976).

59 Beher, W. T. and Anthony, W. L.: Effect of dihydrocholesterol and soy bean sterols on elevated tissue cholesterol. Proc. Soc. exp. Biol. Med. *86:* 589–590 (1954).

60 Beher, W. T. and Anthony, W. L.: Effects of β-sitosterol and ferric choloride on accumulation of cholesterol in mouse liver. Proc. Soc. exp. Biol. Med. *90:* 223–225 (1955).

61 Beher, W. T. and Anthony, W. L.: Liver cholesterol mobilization in mice as effected by dietary β-sitosterol and cholic acid. Proc. Soc. exp. Biol. Med. *99:* 356–358 (1958).

62 Beher, W. T.; Anthony, W. L., and Baker, G. D.: Effect of β-sitosterol on development and regression of cholesterol atherosclerosis in rabbits. Fed. Proc. *15:* 216–217 (1956).

63 Beher, W. T.; Anthony, W. L., and Baker, G. D.: Effects of beta-sitosterol on regression of cholesterol atherosclerosis in rabbits. Circulation Res. *4:* 485–487 (1956).

64 Beher, W. T.; Baker, G. D., and Anthony, W. L.: Effect of dihydrocholesterol and β-sitosterol on cholesterol atherosclerosis in rabbits. Circulation Res. *5:* 202–206 (1957).

65 Beneke, F. W.: Studien über das Vorkommen, die Verbreitung und die Funktion von Gallenbestandteilen in den tierischen und pflanzlichen Organismen (Rickert, Giessen 1862).

66 Beneke, G. M. R.: Cholesterin im Pflanzenreich gefunden. Ann. *122:* 249–255 (1862).

67 Benevista, P.; Hewlins, M. J. E. et Fritig, B.: La biosynthèse des stérols dans les tissus de tabac cultivés *in vitro.* Cinétique de formation des stérols et de leurs précurseurs. Eur. J. Biochem. *9:* 526–533 (1969).

68 Bennett, R. D. and Heftmann, E.: Biosynthesis of stigmasterol from sitosterol in *Digitalis lanata.* Steroids *10:* 403–407 (1969).

69 Bennett, R. D.; Heftmann, E.; Preston, W. H., Jr., and Haun, J. R.: Biosynthesis of sterols and sapogenins in *Dioscora spiculiflora.* Archs Biochem. *103:* 74–83 (1963).

70 Bennett, R. D.; Heftmann, E., and Winter, B. J.: A function of sitosterol. Phytochemistry *8:* 2325–2328 (1969).

71 Berge, K. G.; Achor, R. W. P.; Barker, N. W., and Power, M. H.: A comparison of nicotinic acid, sitosterol, and safflower oil in the treatment of hypercholesterolemia. Circulation *18:* 490–491 (1958).

72 Berge, K. G.; Achor, R. W. P.; Barker, N. W., and Power, M. H.: Comparison of the treatment of hypercholesterolemia with nicotinic acid, sitosterol, and safflower oil. Am. Heart J. *58:* 849–853 (1959).

73 Bergen, S. S., Jr. and Van Itallie, T. B.: Approaches to the treatment of hypercholesterolemia. Ann. intern. Med. *58:* 355–366 (1963).

74 Bergmann, E. D. and Levinson, Z. H.: Fate of beta-sitosterol in housefly larvae. Nature, Lond. *182:* 723–724 (1958).

75 Bergmann, W.: Note on bombicysterol. J. biol. Chem. *107:* 527–532 (1934).

76 Bergmann, W.: The plant sterols. A. Rev. Plant Physiol. *4:* 383–426 (1953).

77 Bergmann, W.: Evolutionary aspects of the sterols; in Cook, Cholesterol, chemistry, biochemistry, and pathology, pp. 435–444 (Academic Press, New York 1958).

78 Bergmann, W.: Sterols: their structure and distribution; in Florkin and Mason, Comparative biochemistry. A comprehensive treatise, vol. III, pp. 103–162 (Academic Press, New York 1962).

79 Berkowitz, D.; Croll, M. W., and Likoff, W.: The effect of sitosterol, nicotinic acid and triparanol on fat tolerance. Am. J. clin. Nutr. *10:* 107–110 (1962).

80 Berkowitz, D.; Likoff, W., and Sklaroff, D. M.: The effect of sitosterol on radioactive fat absorption patterns. Am. J. Cardiol. *4:* 282–286 (1959).

81 Berkowitz, D.; Sklaroff, D., and Likoff, W.: Hypercholesterolemia: considerations in a rational approach to its treatment. Fed. Proc. *19:* 19 (1960).

82 Berkowitz, D.; Sklaroff, D. M.; Spitzer, J. J., and Likoff, W.: Relation of fat tolerance to coronary atherosclerosis. Circulation *22:* 723 (1960).

83 Bernstein, S. and Waillis, E. S.: The structure of β-sitosterol, and its preparation from stigmasterol. J. org. Chem. *2:* 341–345 (1937).

84 Berthaux, P.: Hyperlipemies et hypercholestérolemies. Bull. méd., Paris *71:* 333–342 (1957).

85 Berthold-Godefroy, N. et Wolff, R.: La résorption du cholestérol-4-^{14}C chez le rat et l'influence du sitostérol. C. r. hebd. Séanc. Acad. Sci., Paris *246:* 2825–2828 (1958).

86 Berthold-Godefroy, N. et Wolff, R.: La résorption intestinale du cholestérol-4-^{14}C chez le rat, et l'influence du sitostérol. Revue Athérosclér. *1:* 177–180 (1959).

87 Best, M. M. and Duncan, C. H.: Effects of sitosterol on the cholesterol concentration in serum and liver in hypothyroidism. Circulation *14:* 344–348 (1956).

88 Best, M. M. and Duncan, C. H.: Observation on the mechanism of the hypocholesterolemic effect of sitosterol. Circulation *14:* 911 (1956).

89 Best, M. M. and Duncan, C. H.: Modification of abnormal lipid patterns in atherosclerosis by administration of sitosterol. Ann. intern. Med. *45:* 614–622 (1956).

90 Best, M. M. and Duncan, C. H.: Inhibitory effect of 'isocholesterol' on the absorption of cholesterol. Circulation Res. *5:* 401–404 (1957).

91 Best, M. M. and Duncan, C. H.: Inhibition of cholesterol absorption by wool fat sterols. Clin. Res. *5:* 182 (1957).

92 Best, M. M. and Duncan, C. H.: Factors influencing the hypocholesterolemic effect of sitosterol. Circulation *16:* 861 (1957).

93 Best, M. M. and Duncan, C. H.: Sterol inhibitors of cholesterol absorption. Am. J. Med. *22:* 962 (1957).

94 Best, M. M. and Duncan, C. H.: Effects of the esterification of supplemental cholesterol and sitosterol in the diet. J. Nutr. *65:* 169–181 (1958).

95 Best, M. M. and Duncan, C. H.: Effects of thiouracil and sitosterol on diet-induced hypercholesterolemia and lipomatous arterial lesions in the rat. Am. Heart J. *58:* 214–220 (1959).

96 Best, M. M. and Duncan, C. H.: Effect of cholesterol-lowering durgs on serum triglycerides. J. Am. med. Ass. *187:* 37–40 (1964).

97 Best, M. M. Duncan, C. H.; Van Loon, E. J., and Wathen, J. D.: Lowering of serum cholesterol by the administration of plant sterols. Circulation *10:* 201–206 (1954).

98 Best, M. M.; Duncan, C. H.; Van Loon, E. J., and Wathen, J. D.: The effects of sitosterol on serum lipids. Am. J. Med. *19:* 61–70 (1955).

99 Best, M. M.; Duncan, C. H., and Wathen, J. D.: Effect of sitosterol on the hyperlipemia of myxedema. Circulation *12:* 482–483 (1955).

100 Best, M. M.; Duncan, C. H.; Wathen, J. D., and Kerman, H. D.: Serum lipid fractions

in induced and spontaneous hypothyroidism; modification by beta-sitosterol. Am. J. Med. *19:* 135–136 (1955).

101 Best, M. M.; Duncan, C. H.; Wathen, J. D.; Van Loon, E. J., and Shipley, R. E.: Effects of prolonged administration of sitosterol on serum lipids. Circulation. *10:* 590 (1954).

102 Betzien, G.; Bracharz, H.; Diezel, P. B.; Franke, H.; Kuhn, R. und Seidl, T.: Zur Wirkung des Sitosterins auf den Cholesterin-Stoffwechsel beim Warmblüter. Arznei-mittel-Forsch. *11:* 751–762 (1961).

103 Bevans, M. and Mosbach, E. H.: Biological studies of dihydrocholesterol. Production of biliary concrements and inflammatory lesions of the biliary tract in rabbits. Archs Path. *62:* 112–117 (1956).

104 Beveridge, J. M. R.: Role of fat in nutrition. II. Role of unsaturated fat in adult nutrition. Am. J. publ. Hlth *47:* 1370–1380 (1957).

105 Beveridge, J. M. R. (assigned to Queens University, Kingston, Ont.): Food composition having hypocholesterolemic activity. Can. 567,202 (Dec. 9, 1958).

106 Beveridge, J. M. R. and Connell, W. F.: The effect of commercial margarines on plasma cholesterol levels. Circulation *22:* 674 (1960).

107 Beveridge, J. M. R. and Connell, W. F.: The effect of commercial margarines on plasma cholesterol levels in man. Am. J. clin. Nutr. *10:* 391–397 (1962).

108 Beveridge, J. M. R. and Connell, W. F.: Dietary sitosterol and plasma cholesterol levels in man; in Sissakian, Proc. 5th Int. Congr. of Biochemistry, sect. 12 (12.8), p. 227 (Pergamon Press, Oxford 1963).

109 Beveridge, J. M. R.; Connell, W. F., and Mayer, G.: Further studies on dietary factors affecting plasma lipid levels in humans. Circulation *12:* 499 (1955).

110 Beveridge, J. M. R.; Connell, W. F., and Mayer, G.: The nature of the plasma cholesterol elevating and depressant factors in butter and corn oil. Circulation *14:* 484 (1956).

111 Beveridge, J. M. R.; Connell, W. F., and Mayer, G.: Dietary factors affecting the level of plasma cholesterol in humans: the role of fat. Can. J. Biochem. *34:* 441–455 (1956).

112 Beveridge, J. M. R.; Connell, W. F., and Mayer, G.: Plasma cholesterol depressant factor in corn oil. Fed. Proc. *16:* 11 (1957).

113 Beveridge, J. M. R.; Connell, W. F., and Mayer, G.: The nature of the substances in dietary fat affecting the level of plasma cholesterol in humans. Can. J. Biochem. *35:* 257–270 (1957).

114 Beveridge, J. M. R.; Connell, W. F.; Mayer, G.; Firstbrook, J. B., and De Wolfe, M.: The effect of certain vegetable and animal fats on plasma lipids in humans. Circulation *10:* 593 (1954).

115 Beveridge, J. M. R.; Connell, W. F.; Mayer, G.; Firstbrook, J. B., and De Wolfe, M.: The effects of certain vegetable and animal fats on the plasma lipids of humans. J. Nutr. *56:* 311–320 (1955).

116 Beveridge, J. M. R.; Connell, W. F.; Mayer, G. A., and Haust, H. L.: Further information on the nature of the plasma depressant factor in corn oil. Revue can. Biol. *16:* 465–466 (1957).

117 Beveridge, J. M. R.; Connell, W. F.; Mayer, G. A., and Haust, H. L.: Further assessment of the role of sitosterol in accounting for the plasma cholesterol depressant action of corn oil. Circulation *16:* 491 (1957).

118 Beveridge, J. M. R.; Connell, W. F.; Mayer, G. A., and Haust, H. L.: Influence of

experimental design on assessment of hypocholesterolemic activity of certain vegetable oils. Fed. Proc. *17:* 470 (1958).

119 Beveridge, J. M. R.; Connell, W. F.; Mayer, G. A., and Haust, H. L.: Plant sterols, degree of unsaturation, and hypercholesterolemic action of certain fats. Can. J. Biochem. *36:* 895–911 (1958).

120 Beveridge, J. M. R.; Connell, W. F.; Mayer, G. A., and Haust, H. L.: Response of man to dietary cholesterol. J. Nutr. *71:* 61–65 (1960).

121 Beveridge, J. M. R.; Connell, W. F.; Mayer, G. A., and White, M.: Hypocholesterolemic activity of corn oil fractions in human subjects. Proc. Can. Fed. biol. Soc. *1:* 6–7 (1958).

122 Beveridge, J. M. R.; Haust, H. L., and Connell, W. F.: Magnitude of the hypocholesterolemic effect of dietary sitosterol in man. J. Nutr. *83:* 119–122 (1964).

123 Bhattacharyya, A. K. and Connor, W. E.: Beta-sitosterol and xanthomatosis: a newly decribed lipid storage disease in two sisters. J. clin. Invest. *52:* 9a (1973).

124 Bhattacharyya, A. K. and Connor, W. E.: β-Sitosterolemia and xanthomatosis. A newly described lipid storage disease in two sisters. J. clin. Invest. *53:* 1033–1043 (1974).

125 Bhattacharyya, A. K. and Connor, W. E.: Metabolic studies in the new lipid storage disease β-sitosterolemia and xanthomatosis. Circulation *52:* II-5 (1975).

126 Bhattacharyya, A. K.; Connor, W. E., and Spector, A. A.: Excretion of sterols from the skin of normal and hypercholesterolemic humans. J. clin. Invest. *51:* 2060–2070 (1972).

127 Bieberdorf, F. A. and Wilson, J. D.: Studies on the mechanism of unsaturated fats on cholesterol metabolism in the rabbit. J. clin. Invest. *44:* 1834–1844 (1965).

128 Billings, F. T.: Variability of serum cholesterol in hypercholesterolemia. Archs intern. Med. *110:* 53–56 (1962).

129 Bissett, S. K. and Cook, R. P.: Faecal excretion in man after the administration of mixed plant sterols (sitosterols). Biochem. J. *63:* 13P (1956).

130 Blohm, T. R.; Lerner, L. J.; Kariya, T., and Winje, M. E.: Inhibition of dietary hypercholesterolemia and atherogenesis in the chicken by a new synthetic compound. Circulation Res. *6:* 260–265 (1958).

131 Blomstrand, R. and Ahrens, E. H., Jr.: Absorption of fats studied in a patient with chyluria. III. Cholesterol. J. biol. Chem. *233:* 327–330 (1958).

132 Böhme, H. und Völcker, P. E.: Zur Kenntnis der nichtflüchtigen Anteile des Pomeranzenschalenöls. Arch. Pharm. *292:* 529–536 (1959).

133 Böhle, E.; Harmuth, E. und Rajewsky, M.: Über die Resorption pflanzlicher Sterine beim Menschen. Z. klin. Chem. *2:* 105–114 (1964).

134 Boorman, K. N. and Fisher, H.: The absorption of plant sterols by the fowl. Br. J. Nutr. *20:* 689–701 (1966).

135 Borgström, B.: Absorption of fats. Proc. nutr. Soc. *23:* 34–46 (1967).

136 Borgström, B.: Quantitative aspects of the intestinal absorption and metabolism of cholesterol and β-sitosterol in the rat. J. Lipid Res. *9:* 473–481 (1968).

137 Borgström, B.: Quantification of cholesterol absorption in man by faecal analysis after the feeding of a single isotope-labeled meal. J. Lipid Res. *10:* 331–337 (1969).

138 Borgström, B.: Plant sterols; in Feldman, Nutrition and cardiovascular diseases, pp. 117–138 (Appleton Century Crofts, New York 1976).

139 Boutry, J.-L. et Baron, C.: Etude biochimique des planctons. I. Insaponifiables de

planctons marin et lacustra; stérols d'un plancton. Bull. Soc. Chim. biol. *49:* 157–167 (1967).

140 Boyle, E., Jr.: Types of elevated serum lipid levels in man, and their management. J. Am. Geriat. Soc. *10:* 822–830 (1962).

141 Boyle, E.; Wilkinson, C.F., Jr.; Jackson, R.S., and Benjamin, M.R.: Lipoprotein studies on humans subjected to controlled dietary regimens with subtraction and addition of fats, sitosterol, and dihydrosterol. Circulation *8:* 443–444 (1953).

142 Bragdon, J.H.; Zeller, J.H., and Stevenson, J.W.: Swine and experimental atherosclerosis. Proc. Soc. exp. Biol. Med. *95:* 282–284 (1957).

143 Braus, H.J.; Eck, J.W.; Mueller, W.M., and Miller, F.D.: Isolation and identification of a sterol glucoside from whisky. J. Agric. Food Chem. *5:* 458–459 (1957).

144 Breslaw, L.: Xanthoma tuberosum. A six-month control study. Am. J. Med. *25:* 487–496 (1958).

145 Briones, E.E.; Buscaglia, M.D.; Tyler, N.E., and Kottke, D.A.: Differences in sterol balance in homozygous and heterozygous type IIA hyperlipoproteinemia. Circulation *52:* II–82 (1975).

146 Bronte-Stewart, B.: Lipids and atherosclerosis. Fed. Proc. *20:* 127–134 (1961).

147 Bronte-Stewart, B.; Antonis, A.; Eales, L., and Brock, J.F.: Effects of feeding different fats on serum-cholesterol level. Lancet *i:* 521–526 (1956).

148 Bronte-Stewart, B. and Blackburn, H.: The effect of corn oil on lipid clearance in patients with ischemic heart disease; in Sinclair, Essential fatty acids, pp.180–185 (Academic Press, New York 1958).

149 Brown, H.: Fashioning a practical vegetable-oil food pattern – an experimental study. J. Am. diet. Ass. *38:* 536–539 (1961).

150 Brown, H. and Lewis, L.A.: Effect of high fat and high protein diets on tissue lipids. Circulation *14:* 488 (1956).

151 Brown, H.B. and Page, I.H.: Critical balance between saturated and unsaturated fat necessary to reduce serum cholesterol. Circulation *24:* 1085 (1961).

152 Brown, H.B. and Page, I.H.: Effect of polyunsaturated eggs on serum cholesterol. J. Am. diet. Ass. *46:* 189–192 (1956).

153 Brust, M. and Fraenkel, G.: The nutritional requirements of the larvae of a blowfly, *Phormia regina.* Physiol. Zool. *28:* 186–204 (1955).

154 Buffoni-Nardin, F.: L'azione ostacolante del β-sitosterolo sull'assorbimento intestinale di colesterolo. Archo ital. sci. farm. *9:* 130 (1959).

155 Burián, R.: Über Sitosterin (Ein Beitrag zur Kenntniss des Pflanzensterins.) Sitzber. Akad. Wiss. Wien, Math. Naturw. Klin. Abt. IIb, *106:* 549–572 (1897).

156 Burke, K.A.; McCandless, R.F.J., and Kritchevsky, D.: Effect of soybean sterols on liver deposition of cholesterol-C^{14}. Proc. Soc. exp. Biol. Med. *87:* 87–88 (1954).

157 Cadillos, R.A.; Reeves, R.E., and Swartzwelder, C.: Influence of added cholesterol and dihydrocholesterol upon multiplication of *Entamoeba histolytica* in MS-F medium. Expl Parasit. *11:* 305–310 (1961).

158 Cailleau, R.: L'activité du quelques stérols envisagés comme facteurs de croissance pour la flagellé *Trichomonas columbay.* C. r. Séanc. Soc. Biol. *122:* 1027–1028 (1936).

159 Campagnoli, M.: Systancias reductoras del colesterol. Día Méd. *33:* 1350–1351(1961).

160 Castle, M.; Blondin, G., and Nes, W.R.: Evidence for the origin of the ethyl group of β-sitosterol. J. Am. chem. Soc. *85:* 3306–3308 (1963).

161 Cerbulis, J. and Taylor, N. W.: Neutral lipid and fatty acid composition of earthworms *(Lumbricus terrestris)*. Lipids *4:* 363–368 (1969).

162 Cevallos, W. H.; Holmes, W. L., and Bortz, W. M.: Cholesterol balance studies in normal and hyperlipoproteinemic man. Fed. Proc. *31:* 291 (1972).

163 Chapman, D.: Some studies of lipids, lipid-cholesterol and membrane systems. Studies of lipid-cholesterol interaction; in Chapman and Wallach, Biologic membranes. Physical facts and functions, vol. II, pp. 91–104, 118–128 (Academic Press, New York 1973).

164 Chapman, D.; Owens, N. F.; Phillips, M. C., and Walker, D. A.: Mixed monolayers of phospholipids and cholesterol. Biochim. biophys. Acta *183:* 458–465 (1969).

165 Chattopadhyay, D. P.: Studies on the effect of vegetable oils on the pathogenesis of atherosclerosis in rabbits. Annls Biochem. exp. Med., Calcutta *21:* 55–74 (1961).

166 Chen, J.-S.: The effect of long-term vegetable diet on serum lipids in man. 10th Int. Congr. Gerontology, Jerusalem 1975, abstr., II, p. 55.

167 Chiu, G. C.: Mode of action of cholesterol-lowering agents. Archs intern. Med. *108:* 117–132 (1961).

168 Chiu, G. C.; Shipley, R. E., and Kohlstaedt, K. G.: Correspondence: Are sitosterols safe? Clin. Pharmacol. Ther. *3:* 823–825 (1962).

169 Choquette, G.; David, P. et Drouin-Naud, C.: Athérosclérose, lipides sériques et nutrition. I. Revue de la littérature. Un. méd. Can. *88:* 1390–1400 (1959).

170 Clarenburg, R.; Chung, I. A. K., and Wakefield, L. M.: Reducing the egg cholesterol level by including emulsified sitosterol in standard chicken diet. J. Nutr. *101:* 289–298 (1971).

171 Clark, A. J. and Bloch, K.: Conversion or ergosterol to 22-dehydrocholesterol in *Blatella germanica.* J. biol. Chem. *234:* 2589–2594 (1959).

172 Clarkson, T. B.; King, J. S., Jr., and Warnock, N. H.: The hypocholesterolizing effect of Gallogen and its potentiation by soybean sterols. Circulation Res. *4:* 54–56 (1956).

173 Clarkson, T. B. and Lofland, H. B.: Therapeutic studies on spontaneous atherosclerosis in pigeons. Fed. Proc. *19:* 16 (1960).

174 Clarkson, T. B. and Lofland, H. B.: Therapeutic studies on spontaneous atherosclerosis in pigeons; in Garratini and Paoletti, Drugs affecting lipid metabolism, pp. 314–317 (Elsevier, Amsterdam 1961).

175 Clayton, R. B.: The utilization of sterols by insects. J. Lipid Res. *5:* 3–19 (1964).

176 Clayton, R. B.; Hinkle, P. C.; Smith, D. A., and Edwards, A. M.: The intestinal absorption of cholesterol, its esters and some related sterols and analogues in the roach, *Eurycotis floridana.* Compar. Biochem. Physiol. *11:* 333–350 (1964).

177 Cloetens, W.; Mascart, P.; De Mey, D. et Dernier, J.: Traitement au long cours de l'hypercholestérolémie humaine par le befa-sitostérol. Revue f. Géront. *8:* 136–142 (1962).

178 Cloetens, W.; Mascart, P.; De Mey, D. et Dernier, J.: Action du beta-sitostérol sur la cholesterolémie et le lipidogramme dans la sclérose artérielle. Brux. méd. *44:* 1391–1405 (1964).

179 Cohen, B. I.; Raicht, R. F., and Mosbach, E. H.: Effect of dietary bile acids, cholesterol, and β-sitosterol upon formation of coprostanol and 7-dehydroxylation of bile acids by rat. Lipids *9:* 1024–1029 (1974).

180 Cohen, B. I.; Raicht, R. F.; Shefer, S., and Mosbach, E.: Rate control of sterol metabolism in the rat. Circulation *48:* IV–43 (1973).

181 Cole, A. R. H.: Infrared spectra of natural products; in Zechmeister, Fortschritte der Chemie organischer Naturstoffe, vol. 13, pp. 1–69 (Springer, Berlin 1956).

182 Cole, R. J. and Dutky, S. R.: A sterol requirement in *Turbatrix aceti* and *Panagrellus redivivus.* J. Nematol. *1:* 72–75 (1969).

183 Cole, R. J. and Krusberg, L. R.: Sterol composition of the nematodes *Ditylenchus triformis* and *Ditylenchus dispaci,* and host tissues. Expl Parasit. *21:* 232–239 (1967).

184 Coleman, D. L. and Baumann, C. A.: Intestinal sterols. V. Reduction by intestinal microorganisms. Archs Biochem. Biophys. *72:* 219–225 (1957).

185 Coleman, I. W. and Beveridge, J. M. R.: The effect of dietary fat and the repeated withdrawal of small samples of blood on plasma cholesterol levels in the rat. J. Nutr. *71:* 303–309 (1960).

186 Conner, R. L. and Van Wagtendonk, W. J.: Steroid requirements of *Paramecium aurelia.* J. gen. Microbiol. *12:* 31–36 (1955).

187 Connor, W. E.: Dietary sterols: their relationship to atherosclerosis. J. Am. diet. Ass. *52:* 202–208 (1957).

188 Connor, W. E.: The effects of dietary lipid and sterols on the sterol balance; in Jones, Atherosclerosis, pp. 253–261 (Springer, New York 1970).

189 Connor, W. E.; Rohweider, J. J., and Armstrong, M. L.: Relative failure of saturated fat in the diet to produce atherosclerosis in the rabbit. Circulation Res. *20:* 658–663 (1967).

190 Connor, W. E.; Stone, D. B., and Hodges, R. E.: The interrelated effects of dietary cholesterol and fat upon human serum lipid levels. J. clin. Invest. *43:* 1691–1696 (1964).

191 Connor, W. E.; Witiak, D. T.; Stone, D. B., and Armstrong, M. L.: Cholesterol balance in normal men fed dietary fats of different fatty acid composition. Circulation *36:* II–7 (1967).

192 Connor, W. E.; Witiak, D. T.; Stone, D. B., and Armstrong, M. L.: Cholesterol balance and fecal neutral steroid and bile acid excretion in normal men fed dietary fats of different fatty acid composition. J. clin. Invest. *48:* 1363–1375 (1969).

193 Conrad, L. L. and Furman, R. H.: Observations on the serum lipid and lipoprotein patterns following the administration of sitosterol to normal subjects and patients with hypercholesterolemic nephrosis and primary xanthomatous biliary cirrhosis. Clin. Res. *2:* 127 (1954).

194 Cook, R. P.: The chemistry and biochemistry of the sterols. Proc. Nutr. Soc. *15:* 41–45 (1956).

195 Cook, R. P.: Cholesterol. Chemistry, biochemistry and pathology. Distribution of sterols in organisms and in tissues, pp. 145–180 (Academic Press, New York 1958).

196 Cook, R. P.; Kliman, A., and Fieser, L. F.: The absorption and metabolism of cholesterol and its main companions in the rabbit – with observation of the atherogenic nature of the sterols. Archs Biochem. *52:* 439–450 (1954).

197 Cookson, F. B.; Altschul, R., and Fedoroff, S.: The effect of alfalfa on serum cholesterol and in modifying or preventing cholesterol atherosclerosis in rabbits. J. Atheroscler. Res. *7:* 69–81 (1967).

198 Cookson, F. B. and Fedoroff, S.: Quantitative relationships between administered cholesterol and alfalfa required to prevent hypercholesterolaemia in rabbits. Br. J. exp. Path. *49:* 348–355 (1968).

199 Cooper, E.E.: Dietary and pharmaceutical approaches to atherosclerosis; special reference to beta-sitosterol. Texas St. J. Med. *54:* 29–36 (1958).

200 Corey, J.E.; Dorr, B.B.; Hayes, K.C., and Hegsted, D.M.: Comparative lipid response of four primate species to dietary changes in fat and cholesterol. Fed. Proc. *32:* 315 (1973).

201 Cottet, J.: Est-il possible et utile de modifier le syndrome biochimique de l'athérosclérose? Atti Soc. lombarda sci. med. biol. *11:* 312–329 (1956).

202 Cottet, J.: Peut-on classer pharmacodynamiquement les médicaments modifiant les lipides sanguines? Folia pharmaceut., Istanbul *3:* 235–260 (1956).

203 Cottet, J.: Classification pharmacodynamique des modificateurs de l'hypercholestérolémie. Revue Pharm. Lebanese *4:* 67–84 (1956).

204 Cottet, J.: Les acides gras essentiels en thérapeutiques. Algérie méd. *62:* 47–58 (1958).

205 Cottet, J.: Données récent sur les hypocholestérolemiants. Thérapie *XIII:* XVI–XXXVI (1958).

206 Crombie, W.M.L.: Chemical composition of plant tissues and related data; in Long, Biochemists' handbook, pp. 937–1053 (Spon, London 1961).

207 Cuny, G.; Larcan, A. et Herbeuval, R.: Le béta-sitostérol dans le traitement de l'hypercholestérolémie. Proc. 4th Int. Congr. Gerontology, vol. II, pp. 37–41 (1957).

208 Curran, G.L.: A rational approach to the treatment of atherosclerosis. Am. Practit. *7:* 1412–1417 (1956).

209 Curran, G.L. and Costello, R.L.: Effect of dihydrocholesterol and soybean sterols on cholesterol metabolism in rabbit and rat. Proc. Soc. exp. Biol. Med. *91:* 52–56 (1956).

210 Dam, H.: The formation of coprosterol in the intestine. I. Possible rôle of dihydrocholesterol, and a method of determining dihydrocholesterol in presence of coprosterol. Biochem. J. *28:* 815–819 (1934).

211 Dam, H. und Starup, U.: Über das Schicksal der Pflanzensterine im Tierorganismus. Biochem. Z. *274:* 117–121 (1934).

212 Daskalakis, E.G. and Chaikoff, I.L.: The significance of esterification in the absorption of cholesterol from the intestine. Archs Biochem. *58:* 373–380 (1955).

213 Davignon, J.; Simmonds, W.J., and Ahrens, E.H., Jr.: Usefulness of chromic oxide as an internal standard for balance studies in formula-fed patients and for assessment of colonic function. J. clin. Invest. *47:* 127–138 (1968).

214 Davis, C.B., Jr.; Clancy, R.E.; Cooney, B.E.; Hegsted, D.M., and Hall, J.H.: Effect of mixed fat formula feeding on serum cholesterol level in man. II. Further study utilizing a twenty per cent fat formula. Am. J. clin. Nutr. *8:* 808–811 (1960).

215 Davis, D.L.: Sterol distribution within green and air-cured tobacco. Phytochemistry *11:* 489–494 (1971).

216 Davis, W.W.: Symposium on sitosterol. III. The physical chemistry of cholesterol and β-sitosterol related to the intestinal absorption of cholesterol. Trans. N.Y. Acad. Sci. *18:* ser. II, pp. 123–128 (1955).

217 Day, E.A.; Malcolm, G.T., and Beeler, M.F.: Tumor sterols. Metabolism *18:* 646–651 (1969).

218 Dayton, S. and Pearce, M.L.: Controlled study of the effect of a diet high in unsaturated fat. Circulation *24:* 916 (1961).

219 Dayton, S.; Pearce, M.L.; Hashimoto, S.; Fakler, L.J.; Hiscock, E., and Dixon, W.J.:

A controlled clinical trial of a diet high in unsaturated fat. Preliminary observations. New Engl. J. Med. *206:* 1017–1023 (1962).

220 De Sèze, S.: La feuille d'artichaut *(Cynera scolymus)* en thérapeutique. Presse méd. *2:* 1919–1920, 1923–1924 (1934).

221 De Soldati, L.: Estado actual de los conocimientos sobre etiopatogenia y terapeutica de la arteriosclerosis. Día Méd. *33:* 1658–1688, 1690, 1692 (1961).

222 De Soldati, L.: Alteraciones del metabolismo del colesterol: terapeutica. Revta Asoc. med. Arg. *76:* 39–44 (1962).

223 Den Besten, L.; Connor, W. E., and Kent, T. H.: Effect of cellulose in the diet on the recovery of dietary plant sterols from the feces. J. Lipid Res. *11:* 341–345 (1970).

224 Deuel, H.J., Jr.: The lipids, their chemistry and biochemistry, vol. I, pp. 305–361 (Interscience, New York 1951).

225 Di Bella, S.: Assorbiemento intestinale di steroli. Ol sistema linfatico quale via di assorbiemento degli steroli. Minerva med., Roma *50:* 2269–2273 (1959).

226 Diller, E. R.; Kory, M., and Harvey, O. A.: Effect of fat-free diets and lipid unsaturation on rat tissue cholesterol levels. Proc. Soc. exp. Biol. Med. *108:* 637–640 (1961).

227 Diller, E. R.; Rose, C. L., and Harvey, O. A.: Effect of beta-sitosterol on regression of hyperlipemia and increased plasma coagulability in the chicken. Proc. Soc. exp. Biol. Med. *104:* 173–176 (1960).

228 Diller, E. R.; Woods, B. L., and Harvey, O. A.: Effect of degree of unsaturation of vegetable lipid upon the plasma and hepatic cholesterol content in the rat. Circulation *16:* 505 (1957).

229 Diller, E. R.; Woods, B. L., and Harvey, O. A.: Effect of beta-sitosterol on regression of hypercholesterosis and atherosclerosis in chickens. Proc. Soc. exp. Biol. Med. *98:* 813–817 (1958).

230 Di Paolo: Effetti del β-sitosterolo sul tasso sierico di colesterolo, di lipidi e di acidi grassi poliunsaturi. Archo Pat. Clin. Med. *40:* 211–224 (1963).

231 Dobrowsky, A. und Kohl, A.: Über die Gewinnung des Sitosterins aus Weintrauben-estern und seine Umwandlung in ein D-Vitamin. Chem. Forsch.-Inst. Wirtsch., Öst., Mitt. *7:* 132–134 (1953).

232 Dock, W.; Adlersberg, D.; Eder, H.A.; Kendall, F.E., and Wilkinson, C.F., Jr.: Current concepts in the management of arteriosclerosis (Panel Meet.). Bull. N.Y. Acad. Med. *31:* 198–219 (1955).

233 Doldidze, E.I.: On interrelationship between free and fixed cholesterol in nutrition involving quantitatively different fats (Russian). Vopr. Pitan. *21:* 16–20 (1962).

234 Doorenbos, H.; Valkema, A.J., and Speelman, J.J.: Onderzoek naar de invloed van sitosterol en fenylacetamide (Hypocholesterol) op hypercholesterolemie. Ned. Tijdschr. Geneesk. *102:* 1149–1154 (1958).

235 Dreisbach, R. H.; Brown, B.; Myers, R. B., and Cutting, W. C.: Effect of soy sterols on cholesterol atherosclerosis in rabbits. Fed. Proc. *12:* 317 (1953).

236 Duncan, C.H. and Best, M. M.: Effects of sitosterol on serum lipids of hypercholesterolemic subjects. J. clin. Invest. *34:* 930–931 (1955).

237 Duncan, C. H. and Best, M. M.: Comparative effects of free and esterified sitosterol on serum and liver cholesterol in rats. J. clin. Invest. *35:* 700 (1956).

238 Duncan, C. H. and Best, M. M.: Effects of feeding wool-fat sterols on the sterol content of serum and liver of the rats. J. Nutr. *64:* 425–431 (1958).

239 Duncan, C. H. and Best, M. M.: Long-term use of sitosterol as a hypocholesterolemic agent. J. Kentucky med. Ass. *61:* 45–47 (1963).

240 Dunham, L. W.; Fortner, R. E.; Moore, R. D.; Culp, H. W., and Rice, C. N.: Comparative lymphatic absorption of β-sitosterol and cholesterol by the rat. Archs Biochem. *82:* 50–61 (1959).

241 Dutky, S. R.: An appraisal of the DD 136 nematode for the control of insect populations and some biochemical aspects of its host-parasite relationships. Proc. Joint US-Jap. Seminar Microbial Control of Insect Pests, Fukuoka 1967, pp. 139–140.

242 Dutky, S. R.; Kaplanis, J. N.; Thompson, M. J., and Robbins, W. E.: The isolation and identification of the sterols of the DD-136 insect parasitic nematode and their derivation from its insect hosts. Nematologia *13:* 139–140 (1967).

243 Dutky, S. R.; Robbins, W. E., and Thompson, J. V.: The demonstration of sterols as requirements for the growth, development, and reproduction of the DD-136 nematode. Nematologia *13:* 140 (1967).

244 Earle, N. W.; Lambremont, E. N.; Burks, M. L.; Slatter, B. H., and Bennett, A. F.: Conversion of β-sitosterol to cholesterol in the boll weevil and the inhibition of larval development by two aza sterols. J. econ. Entomol. *60:* 291–293 (1967).

245 Eck, M. et Desbordes, J.: Sur l'hypercholestérinémie exogène et endogène du lapin, influence de la stimulation hépatique. C. r. Séanc. Soc. Biol. *117:* 681–683 (1934).

246 Edwards, P. A. and Green, C.: Incorporation of plant sterols into membranes and its relation to sterol absorption. FEBS Lett. *20:* 97–99 (1972).

247 Elghamry, M. I.: Stimulatory and inhibitory influence of Ladino clover *(Trifolium repens)* phytoestrogens on uterine weight and histology. Zentbl. VetMed. *10 A:* 263–269 (1963).

248 Elghamry, M. I.: Personal correspondence (1976).

249 Ellis, G. W. and Gardner, J. A.: The origin and destiny of cholesterol in the organism. VIII. On the cholesterol content of the liver of rabbits under various diets and during inanition. Proc. R. Soc. Lond. *B 84:* 461–470 (1912).

250 Ellis, L. B.; Blumgart, H. L.; Harken, D. E.; Sise, H. S., and Stare, F. J.: Long-term management of patients with coronary artery disease. Circulation *17:* 945–952 (1958).

251 Emerson, G. A.; Walker, J. B., and Ganaphty, S. N.: Effects of saturated and unsaturated fats and their mixtures on the lipid metabolism of monkeys. J. Nutr. *76:* 6–10 (1962).

252 Eneroth, P.; Hellström, K., and Ryhage, R.: Identification of two neutral metabolites of stigmasterol found in human feces. Steroids *6:* 707–720 (1965).

253 Eneroth, P.; Hellström, K., and Ryhage, R.: Identification and quantification of neutral faecal steroids by gas-liquid chromatography and mass spectrometry: studies of human excretion during two dietary regimens. J. Lipid Res. *5:* 245–262 (1964).

254 Engelberg, H.: Studies of serum cholesterol and low-density lipoproteins, previously lowered by a reduced fat-intake, after the addition of corn oil to the diet. J. chron. Dis. *6:* 229–233 (1957).

255 Engelberg, H.: Effect of safflower oil on serum cholesterol and lipoproteins in patients on a low-fat diet. Geriatrics *13:* 512–516 (1958).

256 Engelhardt, A.: Sitosterin Dilalande – zur Senkung des Cholesterinspiegels. Fortschr. Med. *80:* 519–520 (1962).

257 Enriquez Elesgaray, J.: Tratamiento general de la arterioesclerosis. Arch. Hosp. Univ. Havana *13:* 1–14 (1961).

258 Enselme, M.J.: Biochimie et activité médicamenteuse dans l'athérosclérose. Essai de classification biochimique des corps actifs sur le plaque d'athérosclérose. J. Méd. Lyon *42:* 585–627 (1961).

259 Erickson, B.A.; Coots, R.H.; Mattson, F.H., and Kligman, A.M.: The effect of partial hydrogenation of dietary fats, of the ratio of polyunsaturated to saturated fatty acids, and of dietary cholesterol upon plasma lipids in man. J. clin. Invest. *43:* 2017–2025 (1964).

260 Evans, D.W.; Turner, S.M., and Ghosh, P.: Feasibility of long-term plasma-cholesterol reduction by diet. Lancet *i:* 172–174 (1972).

261 Evrard, E.; Sacquet, E.; Raibaud, P.; Charlier, H.; Dickinson, A.; Eyssen, H., and Hoet, D.P.: Studies on conventional and gnotobiotic rats: effect of intestinal bacteria on fecal lipids and fecal sterols. Ernährungsforschung *10:* 257–263 (1965).

262 Fahrenbach, M.J.; Lewry, H.V.; Riccardi, B.A.; Dunnett, C.W.; Saunders, J.C.; Lourie, E.; Blodinger, J.; Ruegsegger, J.M.; Grant, W.C., and Jukes, T.H.: Effect of β-sitosterol and unsaturated vegetable oil combinations on human serum cholesterol levels. Circulation *18:* 491 (1958).

263 Farquhar, J.W.; Smith, R.E., and Dempsey, M.E.: The effect of β-sitosterol on the serum lipids of young men with atherosclerotic heart disease. Circulation *14:* 77–82 (1956).

264 Farquhar, J.W. and Sokolow, M.: A comparison of the effect of beta-sitosterol and safflower oil, alone and in combination, on serum lipids in humans: a long-term study. Circulation *16:* 494–495 (1957).

265 Farquhar, J.W. and Sokolow, M.: Comparison of the effects of β-sitosterol and safflower oil, alone and in combination, on serum lipids of humans: a long-term study. Circulation *16:* 877 (1957).

266 Farquhar, J.W. and Sokolow, M.: A comparison of the effect of β-sitosterol, gallogen and safflower oil, on serum lipoproteins of humans. Clin. Res. *5:* 36 (1957).

267 Farquhar, J.W. and Sokolow, M.: A comparison of the effect of β-sitosterol, gallogen, and safflower oil on serum lipoproteins of humans. Stanford med. Bull. *15:* 47 (1957).

268 Farquhar, J.W. and Sokolow, M.: Response of serum lipids and lipoproteins of man to beta-sitosterol and safflower oil. Circulation *17:* 890–899 (1958).

269 Favarger, P.: L'absorption intestinale des stérols animaux et végétaux. Bull. Soc. Chim. biol. *40:* 1023–1043 (1958).

270 Favarger, P. et Roth, M.: Influence des stérols sur la résorption des grasses. I. Int. Coll. Biochem. Probl. Lipid, pp.191–200 (1953).

271 Feldman, E.B. and Borgström, N.: Phase distribution of sterols: studies by gel filtration. Biochim. biophys. Acta *125:* 136–147 (1966).

272 Feldman, E.B. and Borgström, N.: Absorption of sterols by intestinal slices *in vitro.* Biochim. biophys. Acta *125:* 148–156 (1966).

273 Fieser, L.F. and Fieser, M.: Advanced organic chemistry, pp.83, T.30.2, 990, T.30.6 (Reinhold, New York 1961).

274 Fink, S.: Studies on hepatic bile obtained from a patient with an external biliary fistula. Its composition and changes after Diamox administration. New Engl. J. Med. *254:* 258–262 (1956).

275 Fisher, H. and Feigenbaum, A. S.: Essential fatty acids of normal and atherosclerotic aortas from chicken receiving differently saturated fats for three years. Nature, Lond. *186:* 85–86 (1960).

276 Fisher, H.; Weiss, H. S., and Griminger, P.: Influence of fatty acids and sterols on atherosclerosis in the avian abdominal aorta. Proc. Soc. exp. Biol. Med. *106:* 61–63 (1961).

277 Fisher, H.; Weiss, H. S., and Griminger, P.: Corn sterols and avian atherosclerosis. Proc. Soc. exp. Biol. Med. *113:* 415–418 (1963).

278 Fisher, H.; Weiss, H. S.; Leveille, G. A.; Feigenbaum, A. S.; Hurwitz, S.; Donis, O., and Lutz, H.: Effect of prolonged feeding of differently saturated fats to laying hens on performance, blood pressure, plasma lipids and changes in the aorta. Br. J. Nutr. *14:* 433–444 (1960).

279 Fishler-Mates, Z.; Budowski, P., and Pinsky, A.: Effect of soy sterols on cholesterol synthesis in the rat. Lipids *8:* 40–42 (1973).

280 Fishler-Mates, Z.; Budowski, P., and Pinsky, A.: Effect of soy sterols on cholesterol metabolism in the rat. Int. J. Vit. Res. *44:* 497–506 (1974).

281 Fomon, S. J. and Bartels, D. J.: Concentrations of cholesterol in serum of infants in relation to diet. J. Dis. Child. *99:* 27–30 (1960).

282 Fraenkel, G.; Reid, J. A., and Blewett, M.: The sterol requirements of the larvae of the beetle *Dermestes vulpinus* Fabre. Biochem. J. *35:* 712–720 (1941).

283 Fredrickson, D. S.: Some biochemical aspects of lipid and lipoprotein metabolism. J. Am. med. Ass. *164:* 1895–1899 (1957).

284 Friedman, M.; Homer, R., and Byers, S. O.: Experimental and clinical study of the effect of β-sitosterol administration on intestinal absorption of dietary cholesterol and the plasma cholesterol content of patients with coronary artery disease. Circulation *12:* 709 (1955).

285 Friedman, M.; Rosenman, R. H., and Byers, S. O.: The effect of beta-sitosterol upon intestinal absorption of cholesterol in the rat. Circulation Res. *4:* 157–161 (1956).

286 Friskey, R. W.; Michaels, G. D., and Kinsell, L. W.: Observations regarding the effects of unsaturated fats. Circulation *12:* 492 (1955).

287 Fröbrich, G.: Untersuchungen über Vitaminbedarf und Wachstumfaktoren bei Insekten. Z. vergl. Physiol. *27:* 335–383 (1939).

288 Fujimoto, G. I. and Jacobson, A. E.: A preparation of β-sitosterol. J. org. Chem. *29:* 3377–3381 (1966).

289 Funch, J. P.; Kristensen, G., and Dam, H.: Effects of various dietary fats on serum cholesterol, liver lipids and tissue pathology in rabbits. Br. J. Nutr. *16:* 497–506 (1962).

290 Furman, R. H.: Hormones, diet and atherosclerosis. Am. Practit. *8:* 741–745 (1957).

291 Furman, R. H. and Robinson, C. W., Jr.: Hypocholesterolemic agents. Med. Clins N. Am. *45:* 935–959 (1961).

292 Galli, A. et Leluc, R.: Métabolism du choléstérol et médicaments hypocholéstérolemiants. Prod. pharm. *14:* 573–582, 634–641 (1959).

293 Gardner, J. A. and Gainsborough, H.: Blood cholesterol studies in biliary and hepatic disease. Q. Jl Med. *23:* 465–483 (1930).

294 Gawienowski, A. and Gibbs, C.: Identification of cholesterol and progesterone in apple seeds. Steroids *12:* 545–550 (1968).

295 Gerolami, A. and Sarles, H.: Letter: beta-sitosterol and chenodeoxycholic acid in the treatment of cholesterol gall stones. Lancet *ii:* 721 (1975).

296 Gerson, T.; Shorland, F.B., Adams, Y.: The effects of corn oil on the amounts of cholesterol and the excretion of sterol in the rat. Biochem. J. *81:* 584–591 (1962).

297 Gerson, T.; Shorland, F.B., and Dunckley, G.G.: Effect of β-sitosterol on cholesterol and lipid metabolism in the rat. Nature, Lond. *200:* 579 (1963).

298 Gerson, T.; Shorland, F.B., and Dunckley, G.G.: The effect of beta-sitosterol on the metabolism of cholesterol and lipids in rats on a low fat diet. Biochem. J. *92:* 385–390 (1964).

299 Gerson, T.; Shorland, F.B., and Dunckley, G.G.: The effect of beta-sitosterol on the metabolism of cholesterol and lipids in rats on a diet containing coconut oil. Biochem. J. *96:* 399–403 (1965).

300 Gey, K.F. and Pletscher, A.: Inability of refined corn oil to influence spontaneous arteriosclerosis of old hens. Nature, Lond. *189:* 491–492 (1961).

301 Giacovazzo, M.; Dal Fabbro, G.; Borso, M.T. e Garufi, L.: Nuovi orientamenti nella terapia dell'aterosclerosi. Rass. Fisiopat. clin. terap. *32:* 433–457 (1960).

302 Giotti, A. e Buffoni, F.: L'azione ostacolante del β-sitosterolo sull'assorbimento intestinale del colesterolo. Boll. soc. ital. Biol. sper. *34:* 1332–1336 (1958).

303 Giotti, A. e Buffoni, F.: L'influenza del β-sitosterolo sulla epatotesaurismosi colesterolica. Archo Sci. Biol., Bologna *43:* 206–230 (1959).

304 Glover, J. and Green, C.: Studies on the absorption and metabolism of sterols: mode of absorption; in Popják and Le Breton, Biochemical problems of lipids, pp. 359–364 (Interscience, New York 1956).

305 Glover, J. and Green, C.: Sterol metabolism. 3. The distribution and transport of sterols across the intestinal mucosa of the guinea pig. Biochem. J. *67:* 308–316 (1957).

306 Glover, J.; Green, C., and Stainer, D.W.: Sterol Metabolism. 5. The uptake of sterols by organelles of intestinal mucosa and the site of their esterification during absorption. Biochem. J. *72:* 82–87 (1959).

307 Glover, J.; Leat, W.M.F., and Morton, R.A.: Sterol metabolism. 2. The absorption and metabolism of (^{14}C) ergosterol in the guinea pig. Biochem. J. *66:* 214–222 (1957).

308 Glover, J. and Morton, R.A.: The absorption and metabolism of sterols. Br. med. Bull. *14:* 226–233 (1958).

309 Glover, J. and Stainer, D.W.: Sterol metabolism. 4. The absorption of 7-dehydrocholesterol in the rat. Biochem. J. *72:* 79–82 (1959).

310 Glover, M.; Glover, J., and Morton, R.A.: Provitamin D$_3$ in tissues and the conversion of cholesterol to 7-dehydrocholesterol *in vivo.* Biochem. J. *51:* 1–9 (1952).

311 Gnudi, A. e Coscelli, C.: Attuali conoscenze sul mecanismo di azione del principali fattori ad attività ipocolesterolemizzante. G. Clin. med. *42:* 1038–1068 (1961).

312 Goad, L.J.: in Pridham, Terpenoids in plants, pp. 159–160 (Academic Press, New York 1967).

313 Goad, L.J.; Gibbons, G.F.; Bolger, L.M.; Rees, H.H., and Goodwin, T.W.: Incorporation [2-^{14}C$_1$-(5R)-5-^{3}H$_1$] mevalonic acid into cholesterol by a rat liver homogenate and into β-sitosterol and 28-isofucosterol by *Larix decidua* leaves. Biol. J. *114:* 885–896 (1969).

314 Goad, L.J. and Goodwin, T.W.: The biosynthesis of sterols in higher plants. Biochem. J. *99:* 735–746 (1966).

315 Goldsmith, G. A.: Highlights on the cholesterol-fats, diets and athrosclerosis problem. J. Am. med. Ass. *176:* 783–790 (1961).

316 Gomez, R. y Prieto, M. E.: Hypercholesterolemia. Revta méd. Chile *90:* 352–360 (1962).

317 Gonzalez, I. E.; Norcia, L. N.; Shetlar, M. R.; Robinson, L. L.; Conrad, L. L., and Furman, R. H.: Canine atherogenesis following I[131] administration and cholesterol feeding. Am. J. Physiol. *197:* 413–422 (1959).

318 Gopalan, C. and Ramanathan, K. S.: Effect of some dietary factors on serum cholesterol. Indian J. med. Res. *46:* 473–481 (1958).

319 Gordan, G. S.; Fitzpatrick, M. E., and Lubich, W. P.: Identification of osteolytic sterols in human breast cancer. Trans. Ass. Am. Physns *50:* 183–188 (1967).

320 Gordon, H. and Brock, J. F.: A practical dietary regime for decreasing the serum-cholesterol level. S. Afr. med. J. *32:* 907–911 (1958).

321 Gordon, H.; Lewis, B., and Brock, J. F.: Effect of different dietary fats on fecal end-products of cholesterol metabolism. Nature, Lond. *180:* 923–924 (1957).

322 Gordon, H.; Lewis, B.; Eales, L., and Brock, J. F.: Dietary fat and cholesterol metabolism. Faecal elimination of bile acids and other lipids. Lancet *273:* 1299–1306 (1957).

323 Gordon, H.; Wilkens, J., and Brock, J. F.: The effect of various dietary factors on the serum-cholesterol level and on the faecal fat content. S. Afr. med. J. *32:* 549–550 (1958).

324 Gordon, S.; Stolzenberg, S. J., and Cekleniak, W. P.: Effects of cholesterol and β-sitosterol in the gerbil. Am. J. Physiol. *197:* 671–673 (1959).

325 Goswanmi, S. K. and Frey, C. F.: Effect of beta-sitosterol on cholesterol-cholic acid induced gallstone formation in mice. Am. J. Gastroent. *65:* 305–310 (1976).

326 Gould, R. G.: Absorbability of dihydrocholesterol and sitosterol. Circulation *10:* 589 (1954).

327 Gould, R. G.: Sterol metabolism and its control; in: Symposium on atherosclerosis, pp. 153–168 (Natn. Academy of Sciences Natn. Research Council, Washington 1955).

328 Gould, R. G.: Symposium on sitosterols. IV. Absorbability of β-sitosterol. Trans. N. Y. Acad. Sci. *18:* ser. II, pp. 129–134 (1955).

329 Gould, G. R. and Cook, R. P.: The metabolism of cholesterol and other sterols in the animal organism; in Cook, Cholesterol, chemistry, biochemistry, and pathology, pp. 237–307 (Academic Press, New York 1958).

330 Gould, R. G.; Jones, R. J.; Le Roy, G. V.; Wissler, R. W., and Taylor, C. B.: Absorbability of β-sitosterol in humans. Metabolism *18:* 652–662 (1969).

331 Gould, R. G.; Lotz, L. V., and Lilly, E. M.: Absorption and metabolic effects of dihydrocholesterol and beta sitosterol. Fed. Proc. *14:* 487 (1955).

332 Gould, R. G.; Lotz, L. V., and Lilly, E. M.: Absorption and metabolism of dihydrocholesterol and beta-sitosterol; in Popják and Le Breton, Biochemical problems in lipids, pp. 353–358 (Interscience, New York 1956).

333 Graham, W. D.; Beare, J. L., and Grice, H. C.: Atheromatous changes in cholesterol-fed rabbits treated with eggplant extract. Circulation Res. *7:* 403–409 (1959).

334 Grande, F.; Anderson, J. T., and Keys, A.: Effect of the unsaponifiable fraction of corn oil on serum cholesterol. Fed. Proc. *17:* 58 (1958).

335 Grande, F.; Anderson, J. T., and Keys, A.: Serum cholesterol in man and the unsaponi-

fiable fraction of corn oil in the diet. Proc. Soc. exp. Biol. Med. *98:* 436–440 (1958).

336 Grande, F. and Wada, S.: The solubility of cholesterol in dietary fats. Fed. Proc. *20:* 96 (1961).

337 Grant, W. C.: Influence of avocado on serum cholesterol. Fed. Proc. *19:* 18 (1960).

338 Grant, W. C.: Influence of avocado on serum cholesterol. Proc. Soc. exp. Biol. Med. *104:* 45–47 (1960).

339 Grasso, S.; Cunning, B.; Imaichi, K.; Michaels, G., and Kinsell, L.: Effects of natural and hydrogenated fats of approximately equal dienoic acid content upon plasma lipids. Metabolism *11:* 920–924 (1962).

340 Gray, M. F.; Lawrie, T. D. V., and Brooks, C. J. W.: Isolation and identification of cholesterol α-oxide and other minor sterols in human serum. Lipids *6:* 836–843 (1971).

341 Grebennik, L. I.; Eroshina, N. V.; Kaidin, D. A.; Yakovleva, A. I., and Shekhnazarova, N. G.: Effect of vegetable sunflower oil on the development of hypercholesteremia and atherosclerosis in rabbits (Russian). Farmakol. Toksik. *25:* 345–350 (1962).

342 Greenberg, S. M.; Herndon, J. F.; Lin, T. H., and Van Loon, E. J.: The antihypercholesteremic effect of essential fatty acids in hypercholesteremic dogs. Am. J. clin. Nutr. *8:* 68–71 (1960).

343 Grossi-Paoletti, E. and Paoletti, P.: Lipids in brain tumors; in Kirsh, Grossi-Paoletti and Paoletti, The experimental biology of human brain tumors, pp. 299–329 (Thomas, Springfield 1972).

344 Grundy, S. M. and Ahrens, E. H., Jr.: The effects of unsaturated dietary fats on absorption, excretion, synthesis, and distribution of cholesterol in man. J. clin. Invest. *49:* 1135–1152 (1970).

345 Grundy, S. M.; Ahrens, E. H., Jr., and Davignon, J.: The interaction of cholesterol absorption and cholesterol synthesis in man. J. Lipid Res. *10:* 304–315 (1969).

346 Grundy, S. M.; Ahrens, E. H., Jr., and Salen, G.: Dietary β-sitosterol as an internal standard to correct for cholesterol losses in sterol balance studes. J. Lipid Res. *9:* 374–389 (1968).

347 Grundy, S. M. and Mok, H. Y.: Effects of low phytosterols on cholesterol absorption in man; in Greten, Lipoprotein metabolism, pp. 112–118 (Springer, Berlin 1976).

348 Grunwald, C.: Effect of sterols on the permeability of alcohol-treated red beet tissue. Plant Physiol. *43:* 484–488 (1968).

349 Grunwald, C.: Effects of free sterols, steryl ester, and steryl glucoside on membrane permeability. Plant Physiol. *48:* 653–655 (1971).

350 Guerra, A. A.: Effect of biologically active substance in the diet on development and reproduction of *Heliothis* spp. J. econ. Entom. *63:* 1518–1521 (1970).

351 Guilleman, P.: L'hypercholestérolémie. Son importance clinique. Sa thérapeutique actuelle. Revue Pratn *11:* 1671–1672, 1675–1676, 1678, 1681–1682 (1961).

352 Gutfinger, T. and Letan, A.: Studies of unsaponifiables in several vegetable oils. Lipids *9:* 658–663 (1974).

353 Haberland, M. E. and Reynolds, J. A.: Self-aggregation of cholesterol in water. Fed. Proc. *32:* 639 (1973).

354 Hanahan, D. J. and Wakil, S. J.: Studies on absorption and metabolism of ergosterol-C^{14}. Archs Biochem. *44:* 150–158 (1953).

355 Hardinge, M. G.; Crooks, H., and Stare, F. J.: Nutritional studies of vegetarians. IV.

Dietary fatty acids and serum cholesterol levels. Am. J. clin. Nutr. *10:* 510–524 (1962).

356 Hardinge, M.G. and Stare, F.J.: Nutritional studies of vegetarians. 2. Dietary and serum levels of cholesterol. Am. J. clin. Nutr. *2:* 83–88 (1954).

357 Hashim, S.A.; Clancy, R.; Hegsted, D.M., and Stare, F.J.: Effect of mixed fat formula feeding on the serum cholesterol level in man. Am. J. clin. Nutr. *7:* 30–34 (1959).

358 Haskins, R.H.; Tulloch, A.P., and Micetich, R.G.: Steroids as the stimulant of sexual reproduction of a species of Phythium. Can. J. Microbiol. *10:* 187–195 (1964).

359 Haust, H.L. and Beveridge, J.M.R.: Type and quality of 3β-hydroxysterols excreted by subjects subsisting on formula rations high in corn oil. J. Nutr. *81:* 13–16 (1963).

360 Heed, W.B. and Kircher, H.W.: Unique sterol in the ecology and nutrition of *Drosophila pachea.* Science *149:* 758–761 (1965).

361 Heftmann, E.: Biochemistry of plant sterols. A. Rev. Plant Physiol. *14:* 225–248 (1963).

362 Heftmann, E.: Biochemistry of steroid saponins and glycoalkaloids. Lloydia *30:* 209–230 (1967).

363 Heftmann, E.: Steroid hormones in plants. Am. Perfum Cosmet. *82:* 47–49 (1967).

364 Heftmann, E.: Biosynthesis of plant steroids. Lloydia *31:* 293–317 (1968).

365 Heftmann, E.: Insect molting hormones in plants. Recent Adv. Phytochem. *3:* 211–227 (1970).

366 Heftmann, E.: Functions of sterols in plants. Lipids *6:* 128–133 (1971).

367 Heftmann, E.: Review. Functions of steroids in plants. Phytochemistry *14:* 891–901 (1975).

368 Heftmann, E.: Steroid hormones in plants. Lloydia *38:* 195–209 (1975).

369 Hegsted, M.: Comparative studies on the effect of dietary fats on serum cholesterol levels. Fed. Proc. *18:* 52–54 (1959).

370 Hegsted, D.M.; Gotsis, A., and Stare, F.J.: The influence of dietary fats on serum cholesterol levels in cholesterol-fed chicks. J. Nutr. *70:* 119–126 (1960).

371 Hellström, K. and Biörck, G.: Terapeutiska möjligheter vid hypercholesterolemia. Nord. Med. *64:* 1392–1393 (1960).

372 Henderson, W.; Reed, W.E.; Steel, G., and Calvin, M.: Isolation and identification of sterols from Pleistocene sediment. Nature, Lond. *231:* 308–310 (1971).

373 Heptinstall, R.H. and Porter, K.A.: The effect of β-sitosterol on cholesterol-induced atheroma in rabbits with high blood pressure. Br. J. exp. Path. *38:* 49–54 (1957).

374 Hermann, B.: Pathogenese, Prophylaxe and Therapie der Atherosklerose. Z. ges. inn. Med. *17:* 45–48, 93–95, 141–144 (1962).

375 Hermann, G.R.: Some experimental studies in hypercholesteremic states. Expl Med. Surg. *5:* 149–159 (1947).

376 Hermann, G.R.: Present day concept of atherosclerosis and various attempts at control with decholesterolizing agents. Arquiv. Brasil Cardiol. *14:* 235–250 (1961).

377 Herrmann, G.R. and Samawi, A.: The effects of various serum cholesterol lowering procedures and agents in patients with coronary artery disease. Tex. Rep. Biol. Med. *20:* 599–614 (1962).

378 Herrmann, R.G.: Effect of taurine, glycine and β-sitosterol on serum and tissue cholesterol in the rat and rabbit. Circulation Res. *7:* 224–227 (1959).

379 Hernandez, H.H. and Chaikoff, I.L.: Purification and properties of pancreatic cholesterol esterase. J. biol. Chem. *228:* 447–457 (1951).

380 Hernandez, H. H. and Chaikoff, I. L.: Do soy sterols interfere with absorption of cholesterol? Proc. Soc. exp. Biol. Med. *87:* 541–544 (1954).

381 Hernandez, H. H.; Chaikoff, I. L.; Dauben, W. G., and Abraham, S.: The absorption of C¹⁴-labeled epicholesterol in the rat. J. biol. Chem. *206:* 757–765 (1954).

382 Hernandez, H. H.; Peterson, D. W.; Chaikoff, I. L., and Dauben, W. G.: Absorption of cholesterol-4-C¹⁴ in rats fed mixed soybean sterols and β-sitosterol. Proc. Soc. exp. Biol. Med. *83:* 498–499 (1953).

383 Herting, D. C.: Annotated bibliography on the effect of phytosterols on cholesterol metabolism and on atherosclerosis. Technical Bull., Biochem. Dpt., Research Labs., Distillation Industries, Div. of Eastman Kodak Co., Rochester, N.Y., Jan. 1, 1957; 1st addendum, Dec. 1, 1958; 2nd addendum, June 1, 1960; 3rd addendum, Dec. 1, 1961; 4th addendum, June 1, 1963.

384 Herting, D. C. and Harris, P. L.: Mixed soybean sterols and cholesterol metabolism. Fed. Proc. *17:* 479 (1958).

385 Herting, D. C. and Harris, P. L.: Effects of water-soluble esters of soybean sterols on cholesterol levels in blood and liver. Fed. Proc. *19:* 18 (1960).

386 Hesse, D.: Über Phytosterin und Cholesterin. Ann. *192:* 175–179 (1878).

387 Hieb, W. F. and Rothstein, M.: Sterol requirements for reproduction of a free-living nematode. Science *160:* 778–779 (1968).

388 Hildreth, E. A.; Mellinkoff, S. M.; Blair, G. W., and Hildreth, D. M.: An experimental study of practical diets to reduce the human serum cholesterol. J. clin. Invest. *30:* 648 (1951).

389 Hidreth, E. A.; Mellinkoff, S. M.; Blair, G. W., and Hildreth, D. M.: The effect of vegetable fat ingestion on human serum cholesterol concentration. Circulation *3:* 641–646 (1951).

390 Hill, E. G.; Silbernick, C. L., and Lundberg, W. O.: Hypercholesterolemic effect of menhaden oil in the presence of dietary cholesterol in swine. Proc. Soc. exp. Biol. Med. *119:* 368–370 (1965).

391 Hiscock, E.; Dayton, S.; Pearce, M. L., and Hashimoto, S.: A palatable diet high in unsaturated fat. J. Am. diet. Ass. *40:* 427–431 (1962).

392 Hochrein, M. und Schleicher, I.: Zum Thema A des vierten Kongresstages: Behandlungsmöglichkeiten bei Arteriosklerose. Kritische Betrachtung zur Entstehung und Behandlung der Arteriosklerose. Med. Klin. *51:* 1691–1707 (1956).

393 Hodges, R. E.; Krehl, W. A., and Stone, D. B.: An effective diet for lowering serum lipids in man. Fed. Proc. *25:* 457 (1966).

394 Hoff, H. R. and De Graff, A. C.: Evaluation of drugs used to reduce serum cholesterol. Am. Heart J. *65:* 132–134 (1963).

395 Hofmann, A. F. and Borgström, B.: Physico-chemical state of lipids in intestinal content during their digestion and absorption. Fed. Proc. *21:* 43–50 (1962).

396 Hofmann, A. F. and Borgström, B.: The intraluminal phase of fat digestion in man: the lipid content of the micellar and oil phases of intestinal content obtained during fat digestion and absorption. J. clin. Invest. *43:* 247–257 (1964).

397 Holman, R. T.: The lipids in relation to atherosclerosis. Am. J. clin. Nutr. *8:* 95–103 (1960).

398 Horlick, L.: Effect of long chain polyunsaturated and saturated fatty acids on the blood lipids in man. Circulation *16:* 491 (1957).

399 Horlick, L.: Studies on the regulation of serum cholesterol levels in man. The effect of

corn oil, ethyl stearate, hydrogenated soybean oil, and nicotinic acid when added to a very low fat basal diet. Lab. Invest. *8:* 723–735 (1959).

400 Horlick, L.: The effect of artificial modification of food on the serum cholesterol level. Can. med. Ass. J. *83:* 1186–1192 (1960).

401 Horlick, L.: Further observation on dietary modifications of serum cholesterol levels. Can. med. Ass. J. *85:* 1127–1131 (1961).

402 Horlick, L.: The mechanism of action of drugs used in the treatment of hyperlipidemia. Pap. Symp. metab. Dis. *1971:* 43–49.

403 Horlick, L.; Cookson, F. B., and Fedoroff, S.: Effect of alfalfa on the excretion of fecal neutral sterols in the rabbit. Circulation *34:* 18 (1967).

404 Horlick, L. and Craig, B. M.: Effect of long-chain polyunsaturated and saturated fatty acids on the serum-lipids of man. Lancet *273:* 566–569 (1957).

405 Hormel Institute: Effects of corn germ components on blood cholesterol Ann. Rep. 10–11 (1957–1958).

406 Horowitz, B. and Winter, G.: A new safflower oil with low iodine value. Nature, Lond. *179:* 582–583 (1957).

407 Howard, A. N.; Gresham, G. A.; Jennings, I. W., and Jones, D.: The effect of drugs on hypercholesterolemia and atherosclerosis induced by semi-synthetic low cholesterol diets; in Kritchevsky, Paoletti and Steinberg, Drugs affecting lipid metabolism, vol. II, pp. 117–127 (Karger, Basel 1967).

408 Huddad, J. G., Jr.; Couranz, S. J., and Avioli, L. V.: Circulating phytosterols in normal females, lactating mothers and breast cancer patients. J. clin. Endocr. *30:* 174–180 (1970).

409 Hudson, J. L.; Diller, E. R.; Pfeiffer, R. R., and Davis, W. W.: Formation of mixed crystals of cholesterol and sitosterol *in vitro* and in rabbit intestine. Proc. Soc. exp. Biol. Med. *102:* 461–463 (1959).

410 Hutchins, R. F. N. and Kaplanis, J. N.: Sterol sulfates in an insect. Steroids *13:* 605–614 (1969).

411 Hutner, S. H. and Holz, G. G.: Lipid requirements of microorganisms. A. Rev. Microbiol. *16:* 189–204 (1962).

412 Ibrahim, M.-B.: Dietetic and pharmacodynamic approach to atherosclerosis. J. Egypt. med. Ass. *46:* 675–688 (1963).

413 Ikekawa, N.; Suzuki, M.; Kobayashi, M., and Tsuda, K.: Studies on the sterol of *Bombyx mori* L. IV. Conversion of the sterol in the silkworm. Chem. pharm. Bull., Tokyo *14:* 834–836 (1966).

414 Ingram, D. S.; Knights, B. A.; McEvoy, I. J., and McKay, P.: Studies in the cruciferae. Changes in composition of the sterol fraction following germination. Phytochemistry *7:* 1241–1245 (1968).

415 Itoh, T.; Tamura, T., and Matsumoto, T.: Sterol composition of 19 vegetable oils. J. Am. Oil Chem. Soc. *50:* 122–125 (1973).

416 Ivy, A. C.; Lin, T.-M., and Karvinen, E.: Absorption of dihydrocholesterol and soya sterols when fed with oleic acid. Am. J. Physiol. *179:* 646–647 (1954).

417 Ivy, A. C.; Lin, T.-M., and Karvinen, E.: Absorption of dihydrocholesterol and soya sterols by the rat's intestine. Am. J. Physiol. *183:* 79–85 (1955).

418 Jacobson, N. L.; Zaletal, J. H., and Allen, R. S.: Effect of various dietary lipids on the blood plasma lipids of dairy calves. J. Dairy Sci. *36:* 832–842 (1953).

419 Jagannathan, S. N.: Effect of feeding fat blends of hydrogenated ground-nut (peanut) fat

and cottonseed oil containing different levels of linoleic acid on serum cholesterol levels in monkeys *(Macaca radiata)* and liver cholesterol concentration in cholesterol-fed rats. J. Nutr. *77:* 317–322 (1962).

420 Jagannathan, S. N.: Effect of feeding different mixtures providing the same amount of linoleic acid, on serum cholesterol levels in monkeys and liver cholesterol concentration in cholesterol-fed rats. J. Nutr. *77:* 323–331 (1962).

421 Jimenez-Diáz, M. C.: Le traitement de l'athérosclérose. Lyon méd. *94:* 1495–1496 (1962).

422 Johnson, D. F.; Bennett, R. D., and Heftmann, E.: Cholesterol in higher plants. Science *140:* 198–199 (1963).

423 Johnson, D. F.; Heftmann, E., and Houghland, G. V. C.: The biosynthesis of sterols in *Solanum tuberosum.* Archs Biochem. *104:* 102–105 (1964).

424 Joliffe, N.: Dietary factors regulating serum cholesterol. Metabolism *10:* 497–513 (1961).

425 Jones, R. J.: Diet and drugs in atherosclerosis. Am. Practit. *9:* 371–375 (1958).

426 Jones, R. J. and Keough, T. F.: Factors in the hypocholesteremic response of patients given a brain extract. J. Lab. clin. Med. *52:* 667–679 (1958).

427 Jones, R. J.; Keough, T. F.; Cummings, D., and Kraft, S.: Factors determining the hypocholesteremic response in patients fed a brain extract. Circulation *16:* 497–498 (1957).

428 Jones, R. J.; Reiss, O. K., and Balter, E. L.: Observations on the mode of action of a hypocholesteremic agent. Circulation *12:* 727–728 (1955).

429 Jones, R. J.; Reiss, O. K., and Huffman, S.: Corn oil and hypocholesteremic response in the cholesterol-fed chick. Proc. Soc. exp. Biol. Med. *93:* 88–91 (1956).

430 Jordan, G. L., Jr.; Chapman, D. M.; De Bakey, M. E., and Halpert, B.: The effect of linoleic and linolenic acid, estradiol valerate and beta-sitosterol upon experimentally induced hypercholesterolemia and the development of lipid changes in the aorta of dogs. Am. J. med. Sci. *241:* 710–717 (1961).

431 Editorial: Hypercholesterolemia and atherosclerosis. J. Am. med. Ass. *160:* 670–671 (1956).

432 Questions and Answers: Hypercholesterolemia. J. Am. med. Ass. *164:* 358 (1957).

433 Editorial: Monomachy with cholesterol. J. Am. med. Ass. *174:* 66 (1960).

434 Joyner, C. and Kuo, P. T.: The serum cholesterol level and lipoprotein pattern following the administration of plant sterols. Circulation *10:* 589 (1954).

435 Joyner, C. and Kuo, P. T.: The effect of sitosterol administration upon the serum cholesterol level and lipoprotein pattern. Am. J. med. Sci. *230:* 636–647 (1955).

436 Jürgens, J. L. and Achor, R. W. P.: Essential hypercholesteremia and its management. Proc. Mayo Clin. *34:* 533–542 (1959).

437 Kägi, P. und Koller, F.: Die Beeinflussung der Serumlipoide durch Sitosterin. Helv. med. Acta *24:* 392–397 (1957).

438 Kahn, D. R. and Munitz, A. J.: Observations on an effect of beta-sitosterol on 'Stypvén clotting time'. Univ. Mich. med. Bull. *26:* 121–126 (1960).

439 Kakis, G. and Kuksis, A.: Exclusion of β-sitosterol from subcellular sites of rat liver penetrated by cholesterol. AOCS, 50th Annu. Meet., abstr. No. 34 (1976).

440 Kalliomäki, J. L.; Waris, E. K., and Vasama, R.: Sitosterolermas verkam på blodets kolesterol och lipoproteiner. Nord. Med. *57:* 460–563 (1957).

441 Kanazawa, A.; Tanaka, N.; Teshima, S., and Kashiwada, K.: Nutritional requirements

of prawn. II. Requirements for sterols (Japanese). Nippon Suisangaku Kaishi (Bull. Jap. Soc. Sci. Fish.) *37:* 211–215 (1971).

442 Kaneda: cited by Reiner et al. [686].

443 Kaplanis, J. N.; Monroe, R. E.; Robbins, W. E., and Louloudes, S. J.: The fate of dietary H^3-β-sitosterol in the adult house fly. Ann. entomol. Soc. Am. *56:* 198–201 (1963).

444 Kaplanis, J. N.; Robbins, W. E.; Monroe, R. E.; Shortino, T. J., and Thompson, M. J.: The utilization and fate of β-sitosterol in the larvae of the housefly, *Musca domestica* L. J. Ins. Physiol. *11:* 251–258 (1965).

445 Karvinen, E.; Lin, T.-M., and Ivy, A. C.: Capacity of human intestine to absorb exogenous cholesterol. J. appl. Physiol. *11:* 143–147 (1957).

446 Kats, K. N.: Therapeutic effect of β-sitosterol in atherosclerosis (literature survey) (Russian). Klin. Med. *40:* 25–27 (1966).

447 Katsui, G.: Resources of vitamin D. I. Sterols in Penicillium molds and the sediment of saké (Japanese). Hakkŏ Kôgaku Zasshi (J. Ferment. Technol. Jap.) *28:* 33–36 (1950).

448 Katz, L. N.: The present status of the management of atherosclerosis. Trans. med. Soc. Lond. *77:* 16–30 (1961).

449 Katz, L. N.: Atherosclerosis. Present status of the management of the disease. Calif. Med. *96:* 373–380 (1962).

450 Katz, L. N. and Pick, R.: The present status of the management of atherosclerosis. Cardiol. prat. *13:* 54–71 (1962).

451 Katz, L. N.; Stamler, J., and Pick, R.: Nutrition and atherosclerosis, pp. 18, 50–60, 69–70, 97–98 (Lea & Febiger, Philadelphia 1958).

452 Katz, M.; Bartov, I.; Budowski, P., and Bondi, A.: Inhibition of cholesterol deposition in livers of mice fed phytosterols in short-term experiments J. Nutr. *100:* 1141–1147 (1970).

453 Kaufmann, J. A.: Discussion of Berkowitz, Spitzer, Sklaroff and Likoff, Radioactive fat tolerance studies in coronary artery disease, Sth. med. J., Nashville *53:* 468–472 (1960).

454 Kemp, R. J.; Goad, L. J., and Mercer, E. I.: Sterol composition in maize during germination. Biochem. J. *103:* 53–54 P (1967).

455 Kemp, R. J.; Good, L. J., and Mercer, E. I.: Changes in the levels of composition of the esterified and unesterified sterols of maize seedlings during germination. Phytochemistry *6:* 1609–1615 (1967).

456 Kemp, R. J. and Mercer, E. I.: The sterol esters of maize seedlings. Biochem. J. *110:* 111–118 (1968).

457 Kemp, R. J. and Mercer, E. I.: Studies on the sterols and sterol esters of the intracellular organelles of maize shoots. Biochem. J. *110:* 119–125 (1968).

458 Kesten, H. D. and Silbowitz, R.: Experimental atherosclerosis and soya lecithin. Proc. Soc. exp. Biol. Med. *49:* 71–73 (1942).

459 Keys, A. and Anderson, J. T.: The relationship of the diet to the development of atherosclerosis in man. Symp. on Atherosclerosis, pp. 181–197 (Natn. Academy of Science Natn. Research Council, Washington 1955).

460 Keys, A.; Anderson, J. T., and Grande, F.: 'Essential' fatty acids, degree of unsaturation and effect of corn (maize) oil on the serum-cholesterol level in man. Lancet *272:* 66–68 (1957).

461 Keys, A.; Anderson, J.T., and Grande, F.: Fats and disease. Lancet *272:* 992–993 (1957).

462 Keys, A.; Anderson, J.T., and Grande, F.: Prediction of serum-cholesterol responses of man to changes in fats in the diet. Lancet *273:* 959–966 (1957).

463 Keys, A.; Anderson, J.T., and Grande, F.: Serum cholesterol in man: diet fat and intrinsic responsiveness. Circulation *19:* 201–214 (1959).

464 Khaletskii, A.M.: Studies in the field of β-sitosterol and non-steroid androgens (Russian). Med. Promyshl. SSSR *11:* 32–37 (1961).

465 Khaletskii, A.M.: β-Sitosterol (from kraft papermaking) (Russian). Bumazh. Promyshl. *36:* 10–11 (1961).

466 Kimmerling, H.W.: Concepts, theory, and treatment of atherosclerosis. J. Am. Geriat. Soc. *10:* 865–876 (1962).

467 King, J.S., Jr.; Clarkson, T.B., and Warnock, N.H.: The hypocholesterolizing effects of drugs – fed singly or in combination – on cholesterol-fed cockerels. Circulation Res. *4:* 162–166 (1956).

468 Kingsbury, K.J. and Morgan, D.M.: The possible effect of lipaemia on the postprandial increases of the plasma cholesterol and phospholipid. Q. Jl exp. Physiol. *45:* 35–39 (1960).

469 Kinsell, L.W.: Reduction of serum cholesterol with the use of diets high in vegetable fat. Mod. Nutr. *7:* 6–8 (1954).

470 Kinsell, L.W.: Effects of high-fat diets on serum lipids. Animal vs. vegetable fats. J. Am. diet. Ass. *30:* 685–688 (1954).

471 Kinsell, L.W.: Nutrition in vascular disease. Borden's Rev. Nutr. Res. *17:* 1–9 (1956).

472 Kinsell, L.W.: Symposium on significance of lowered cholesterol levels. J. Am. med. Ass. *170:* 2201–2202 (1959).

473 Kinsell, L.W.: Some thoughts regarding the P:S ratio concept. Am. J. clin. Nutr. *12:* 228–229 (1963).

474 Kinsell, L.W.; Cochrane, G.C.; Balch, H.E., and Foreman, N.: The effects of specific steroids and phosphatides upon the levels of plasma lipids. Circulation *8:* 450 (1953).

475 Kinsell, L.W.; Cochrane, G.C.; Smyrl, S.; Fukayama, G., and Coelho, M.: Further experience with high vegetable-fat diet in patients with extensive diabetic vascular disease. Circulation *10:* 593 (1954).

476 Kinsell, L.W. and Michaels, G.D.: Hormonal-nutritional – lipid relationships. Fed. Proc. *14:* 661–666 (1955).

477 Kinsell, L.W. and Michaels, G.D.: Letter to editor. In reply to comments in Nutrition Reviews. Am. J. clin. Nutr. *3:* 247–253 (1955).

478 Kinsell. L.W.; Michaels, G.D.; Cochrane, G.C.; Partridge, J.W.; Jahn, J.P., and Balch, H.E.: Effect of vegetable fat on hypercholesterolemia and hyperphospholipidemia. Observations on diabetic and nondiabetic subjects given diets high in vegetable fat and protein. Diabetes *3:* 113–119 (1954).

479 Kinsell, L.W.; Michaels, G.; De Wind, L.; Partridge, J., and Boling, L.: Serum lipids in normal and abnormal subjects. Observations on controlled experiments. Calif. Med. *78:* 5–10 (1953).

480 Kinsell, L.W.; Michaels, G.D.; Friskey, R.W., and Splitter, S.: Essential fatty acids, lipid metabolism and vascular disease; in Sinclair, Essential fatty acids, pp. 125–146 (Academic Press, New York 1958).

481 Kinsell, L. W.; Michaels, G. D.; Partridge, J. W.; Boling, L. A.; Balch, H. E., and Cochrane, G. C.: Effect upon serum cholesterol and phospholipids of diets containing large amount of vegetable fat. Am. J. clin. Nutr. *1:* 224–231 (1953).

482 Kinsell, L. W.; Michaels, G. D.; Wheeler, P.; Flynn, P. F., and Walker, G.: Essential fatty acids and the problem of atherosclerosis. Am. J. clin. Nutr. *6:* 628–631 (1958).

483 Kinsell, L. W.; Partridge, J.; Boling, L.; Margen, S., and Michaels, G.: Dietary modifications of serum cholesterol and phospholipid levels. J. clin. Endocr. *12:* 909–913 (1952).

484 Kinsell, L. W. and Sinclair, H. M.: Fats and disease. Lancet *272:* 883–884 (1957).

485 Klein, L.: Aterosclerosis y colesterolemia. Sem. Méd. *70:* 1871–1874 (1963).

486 Knights, B. A.: Identification of the sterols of oat seeds. Phytochemistry *4:* 857–862 (1965).

487 Konlande, J. E. and Fisher, H.: Evidence for an extra-absorptive, antihypercholesterolemic effect of soy sterols in the chick. Fed Proc. *28:* 515 (1969).

488 Konlande, J. E. and Fisher, H.: Evidence for a nonabsorptive antihypercholesterolemic action of phytosterols in the chicken. J. Nutr. *98:* 435–442 (1969).

489 Korányi, A. and Jáky, M.: Arteriosclerosis és zsíranyagcsere. Diétás vizsgálatok arteriosclerosis betegeknél. Orvosi Szemle *100:* 1828–1834 (1959).

490 Korytek, W. and Metzler, E. A.: Composition of lipids of lima beans and certain other beans. J. Sci. Food. Agric. *14:* 841–844 (1963).

491 Korzenowsky, M.; Walters, C. P.; Harvey, O. A., and Diller, E. R.: Some factors which influence the catalytic activity of pancreatic cholesterol esterase. Proc. Soc. exp. Biol. Med. *105:* 303–305 (1960).

492 Kothari, H. V. and Kritchevsky, D.: Purification and properties of aortic cholesteryl ester hydrolase. Lipids *10:* 322–330 (1975).

493 Kottke, B. A.; Stames, C. S.; Carlo, I. A., and Subbiah, M. T. R.: Cholesterol balance in hyperlipoproteinemia. Circulation *48:* IV–42 (1973).

494 Kottke, B. A. and Subbiah, M. T. R.: Sterol balance studies in patients on solid diets: comparison of two 'nonabsorbable' markers. J. Lab. clin. Med. *78:* 811 (1971).

495 Kottke, B. A. and Subbiah, M. T. R.: Sterol balance studies in patients on solid diets: comparison of two 'nonabsorbable' markers. J. Lab. clin. Med. *80:* 530–538 (1972).

496 Krause, S.: Coronary artery disease; present concepts. Am. J. Cardiol. *1:* 334–339 (1958).

497 Kritchevsky, D. and Defendi, V.: Persistence of sterols other than cholesterol in chicken tissues. Nature, Lond. *192:* 71 (1961).

498 Kritchevsky, D. and Defendi, V.: Deposition of tritium-labeled sterols (cholesterol, sitosterol, lanosterol) in brain and other organs of the growing chicken. J. Neurochem. *9:* 421–425 (1962).

499 Kritchevsky, D. and McCandless, R. F. J.: Weekly variations in serum cholesterol levels of monkeys. Proc. Soc. exp. Biol. Med. *95:* 152–154 (1957).

500 Kritchevsky, D.; Staple, E., and Whitehouse, M. W.: Oxidation of ergosterol by rat and mouse liver mitochondria. Proc. Soc. exp. Biol. Med. *106:* 704–708 (1961).

501 Kritchevsky, D. and Tepper, S. A.: Solubility of cholesterol in various fats and oils. Proc. Soc. exp. Biol. Med. *116:* 104–107 (1964).

502 Kritchevsky, D. and Tepper, S. A.: Cholesterol vehicle in experimental atherosclerosis. IX. Comparison of heated corn oil and heated olive oil. J. Atheroscler. Res. *7:* 647–651 (1967).

503 Kritchevsky, D.; Tepper, S.A.; Williams, D.E., and Story, J.A.: Experimental athero-
     sclerosis in rabbits fed cholesterol-free diets. 7. Interaction of animal and vegetable
     protein with fiber. Atherosclerosis 26: 397–403 (1977).

504 Kritchevsky, D.; Tepper, S.A., and Kothari, H.V.: Effect of hypocholesteremic drugs
     on aortic cholesterol esterase in cholesterol-fed rabbits. Artery 1: 437–444 (1975).

505 Kritchevsky, D.; Tepper, S.A., and Langan, J.: Cholesterol vehicle in experimental
     atherosclerosis. IV. Influence of heated fat and fatty acids. J. Atheroscler. Res. 2:
     115–122 (1962).

506 Kritchevsky, D.; Tepper, S.A., and Story, J.A.: Influence of an eggplant (Solanum
     melongena) preparation on cholesterol metabolism in rats. Exp. Path. 10: 180–183
     (1975).

507 Kritchevsky, D.; Tepper, S.; Vesselinovitch, D., and Wissler, R.W.: Cholesterol vehicle
     in experimental atherosclerosis. 11. Peanut oil. Atherosclerosis 14: 53–64 (1971).

508 Kritchevsky, D.; Winter, P.A.D., and Davidson, L.M.; Cholesterol absorption in
     primates as determined by the Zilversmit isotope ratio method. Proc. Soc. exp. Biol.
     Med. 147: 464–466 (1974).

509 Krivoruchenko, I.V.: Blood lipids and vascular reactivity in atherosclerotic patients.
     Tr. Gos. Inst. Inst. Usoverch, Vrachei im S.M. Kirova No.27: 43–52 (Russian) (1961).
     Ref. Zh. khim.; biol khim., Abstr. No.9S1271 (Russian) (1961). Chem. Abstr. 57: 10450
     G (1962).

510 Krivoruchenko, I.V.: Effect of β-sitosterol on the lipid metabolism and course of
     atherosclerosis (Russian). Sov. Med. 25: 32–38 (1962).

511 Kudchodkar, B.J.; Horlick, L., and Sodhi, H.S.: Effects of plant sterols on cholesterol
     metabolism in man. Atherosclerosis 23: 239–248 (1976).

512 Kudchodkar, B.J.; Sodhi, H.S., and Horlick, L.: Lack of degradation of dietary sterols
     in man. Fed. Proc. 30: 481 (1971).

513 Kudchodkar, B.J.; Sodhi, H.S., and Horlick, L.: Absorption of dietary cholesterol in
     man. Circulation 44: II-3 (1971).

514 Kuksis, A.; Ali, S.S., and Beveridge, J.M.R.: Effect of plant sterol on lipid excretion in
     man. Fed. Proc. 24: 263 (1965).

515 Kuksis, A. and Huang, T.C.: Differential absorption of plant sterols in the dog. Fed.
     Proc. 21: 259 (1962).

516 Kuksis, A. and Huang, T.C.: Differential absorption of plant sterols in the dog. Can. J.
     Biochem. 40: 1493–1504 (1962).

517 Kuksis, A. and Huang, T.C.: A gas-chromatographic study of the in vivo conversion of
     plant sterols to derivatives of coprostanol. Fed. Proc. 23: 553 (1964).

518 Kuksis, A. and Kakis, G.: Lack of effect of plant sterols on feedback control of cho-
     lesterol biosynthesis. Fed. Proc. 32: 238 (1973).

519 Kummerow, F.A.: The relation of adequate nutrition to atherosclerosis. J. Dairy Sci.
     38: 1403–1404 (1955).

520 Kummerow, F.A.: Role of butterfat in nutrition and in atherosclerosis. A review. J.
     Dairy Sci. 40: 1350–1359 (1957).

521 Kuo, P.T.: Effects of lipemia on tissue oxygenation in arteriosclerotic patients. Cir-
     culation 14: 964 (1956).

522 Kuo, P.T. and Joyner, C.R., Jr.: The effect of low fat diet and sitosterol on the serum
     lipids of patients with coronary and peripheral arterial disease. Circulation 14: 499
     (1956).

523 Kúthy: cited by Verzár and McDougall, Absorption from the intestine, p. 214 (Hafner Publishing, New York 1967).

524 Lafille, C.: Editorial. Trait de l'hypercholestérolémie. Prod. Pharm. *17:* 153–154 (1962).

525 Lambert, G.F.; Miller, J.P.; Olsen, R.T., and Frost, D.V.: Hypercholesteremia and atherosclerosis induced in rabbits by purified high fat rations devoid of cholesterol. Proc. Soc. exp. Biol. Med. *97:* 544–549 (1958).

526 Lamberton, J.-N.: Une thérapeutique 'anti-sclérose' de la sclérodermie. L'insaponifiable des huiles d'avocat et de soja. Cinquante applications cliniques du traitement de H. Thiers. Presse méd. *78:* 1235–1236 (1970).

527 Leckert, J.T.; Browne, D.C.; McHardy, G., and Cradic, H.E.: The result of treatment of hypercholesterolemia associated with atherosclerosis with sitosterol and safflower oil derivatives. J. Louisiana St. med. Soc. *110:* 260–266 (1958).

528 Lederer, E.: The origin and function of some methyl groups in branched-chain fatty acids, plant sterols and quinones. Biochem. J. *93:* 449–468 (1964).

529 Lees, R.S. and Lees, A.M.: Effects of sitosterol therapy on plasma lipid and lipoprotein concentration; in Greten, Lipoprotein metabolism, pp.119–124 (Springer, Berlin 1976).

530 Lees, R.S. and Wilson, D.E.: The treatment of hyperlipidemia. New Engl. J. Med. *284:* 186–195 (1971).

531 Lehmann, J.H.: Clinical experience with beta sitosterol, a new anticholesterolemic agent. NW. Med., Seattle *56:* 43–46 (1957).

532 Lehmann, J.H. and Bennett, B.M.: Effect of sitosterol on survival and recurrence rates in moycardial infarction. Circulation *18:* 747–748 (1958).

533 Leonard, J.C.: Hereditary hypercholesterolaemic xanthomatosis. Lancet *271:* 1239–1243 (1956).

534 Lesesne, J.M.; Castor, C.W., and Hoobler, S.W.: Prolonged reduction in human blood cholesterol levels induced by plant sterols. Univ. Mich. med. Bull. *21:* 13–17 (1955).

535 Leveille, G.A. and Fisher, H.: Observations on lipid utilization in hens fed vegetable and animal fat supplemented diets. Poultry Sci. *37:* 658–664 (1958).

536 Lever, W.F.: Primary hypercholesteremia and idiopathic hyperlipemia: clinical, biochemical, and therapeutic considerations. Sth. med. J., Nashville *53:* 454–464 (1960).

537 Levere, A.H.; Bozian, R.C.; Craft, G.; Jackson, R.S., and Wilkinson, C.F., Jr.: The 'sitosterols': variability of serum cholesterol levels and difficulty of evaluating decholesterolizing agents. Metabolism *7:* 338–348 (1958).

538 Levere, A.H.; Bozian, R.C.; Jackson, R.S., and Wilkinson, C.F., Jr.: Analysis of some variables in studies with serum-cholesterol-lowering agents. Circulation *12:* 483 (1955).

539 Levkoff, A.H. and Knode, K.T.: The treatment of familial hypercholesterolemia with a plant sterol. Pediatrics, Springfield *19:* 88–90 (1957).

540 Levy, R.I.; Morganroth, J., and Rifkind, B.M.: Treatment of hyperlipidemia. New Engl. J. Med. *290:* 1295–1301 (1974).

541 Lewis, B.: Effect of certain dietary oils on bile-acid secretion and serum-cholesterol. Lancet *i:* 1090–1092 (1958).

542 Libert, O. and Rogg-Effront, C.: Experimental atherosclerosis and hyperlipidemia in

rats and rabbits. Influence of some alimentary fats. J. Atheroscler. Res. *2:* 186–198 (1962).

543 Lin. L.-L. Wu, C.-Y.; Hsiu, H.-C.; Wang, M.-T., and Chuang, H.: Studies on the diabetes mellitus. 1. The hypoglycemic activity of phytosterin on the alloxan diabetic rat. J. Formosan med. Ass. *66:* 58–65 (1967).

544 Lin, T.-M.: Intestinal absorption of cholesterol. Diss. Abstr. *14:* 2381–2382 (1954).

545 Linde, H.; Ergenc, N. und Meyer, K.: Zur Frage der Existenz von 'γ-Sitosterol'. Nachweis von Cholesterol als Bestandteil des 'γ-Sitosterol' einer Digitalis-Art. Helv. chim. Acta *49:* 1246–1249 (1966).

546 Little, A.: Nutritional factors in atherosclerotic diseases. J. Can. diet. Ass. *19:* 43–49 (1957).

547 Llamas, R.: Modificaciones de la colesterolemia mediante agentes hormonales y far-macológicos. Revta invest. clín. *14:* 157–168 (1962).

548 Lohmann, T.G.; Romack, F.E., and Kummerow, F.A.: Comparison of corn oil and hydrogenated soybean oil on atherosclerosis in mature swine. Fed. Proc. *32:* 316 (1973).

549 Lovelock, J.E.: Discussion to Kinseli et al. [480].

550 Lown, B.; Portman, O.W., and Stare, F.J.: Some comments on the use of agents which lower serum cholesterol. Clin. Pharmacol. Ther. *3:* 421–424 (1962).

551 Lown, B.; Portman, O.W., and Stare, F.J.: Correspondence. Are sitosterols safe? Clin. Pharmacol. Ther. *3:* 825 (1962).

552 Ludwiczak, R.S. and Szczawińska, K.: Składniki obojetne buka zwyczajnego *Fagus silvatica* L. I. β-Sitosterol i betulina z kory bůku. Rocznik chemii *39:* 583–588 (1965).

553 Malmros, H.: The effect of dietary fats on serum cholesterol; in Sinclair, Essential fatty acids, pp.150–155 (Academic Press, New York 1958).

554 Malmros, H. and Wigand, G.: Treatment of hypercholesteremia. Minn. Med. *38:* 864–870 (1955).

555 Malmros, H. and Wigand, G.: The effect on serum-cholesterol of diets containing different fats. Lancet *273:* 1–8 (1957).

556 Mancini, M.; Oriente, P. e D'Andrea, L.: Impiego terapeutico dell'acido 1-4-dicaffeil-chimico principio attivo del carciofo. Minerva med., Roma *51:* 2460–2463 (1960).

557 Marker, R.E. and Shabica, A.C.: Sterols. XCIX. Sterols from various sources. J. Am. chem. Soc. *62:* 2523–2525 (1940).

558 Mathivat, A.: L'hypercholestérolémie. Signification pratique. Conduite à tenir. Prog. méd., Paris *90:* 289–291, 293–294 (1962).

559 Matsui, K.; Akagi, S., and Ugai, F.: Sterol constituents of the alkaloid-free fraction of opium. Chem. pharm. Bull., Tokyo *10:* 872–875 (1962).

560 Matthews, W.S. and Smith, L.L.: Sterol metabolism. III. Sterols of marine waters. Lipids *3:* 239–246 (1968).

561 Mattson, F.H.; Volpenheim, R.A., and Erickson, B.A.: Effect of plant sterol esters on the absorption of dietary cholesterol. J. Nutr. *107:* 1139–1146 (1977).

562 Mayer, G.A.; Connell, W.F., and Beveridge, J.M.R.: Effects of homogeneous formula diets on the clotting time of whole blood. Circulation *16:* 511 (1957).

563 Mayer, G.A.; Connell, W.F.; De Wolfe, M.S., and Beveridge, J.M.R.: Diet and plasma cholesterol levels. Am. J. clin. Nutr. *2:* 316–322 (1954).

564 McCabe, E.S.: A multi-faceted approach towards lowering blood cholesterol. Dela-ware med. J. *34:* 71–74 (1962).

565 McCann, M. B.; Trulson, M. F.; Waddell, W. R.; Dalrymple, W., and Stare, F. J.: The effects of various vegetable oils on the serum lipids of adult American males. Am. J. clin. Nutr. *7:* 35–42 (1959).

566 McCullagh, E. P. and Lewis, L. A.: A study of diet, blood lipids and vascular disease in Trappist monks. New Engl. med. J. *263:* 559–574 (1960).

567 McMillan, G. C.; Weigensberg, B. I., and Ritchie, A. C.: Effects of dietary fats in rabbits fed cholesterol. Severity of atherosclerosis in rabbits fed highly saturated and unsaturated fats. Archs Path. *70:* 220–225 (1960).

568 McOsker, D. E.; Mattson, F. H.; Sweringen, H. B., and Kligman, A. M.: The influence of partially hydrogenated fats on serum cholesterol levels. J. Am. med. Ass. *180:* 380–385 (1962).

569 Mead, J. F.: Fat absorption. Am. J. clin. Nutr. *6:* 606–608 (1958).

570 Mead, J. F.: Discussion to: Kinsell et al. [480].

571 Editorial: Beeinflussung von Cholesterinstoffwechsel und Gefässsklerose durch Beta-Sitosterol (Sammelreferat). Med. Welt, Berlin *50:* 2704–2708 (1962).

572 Mellies, M.; Glück, C. J.; Sweeney, G., and Ishikawa, T.: Plasma phytosterols in children on cholesterol poor, phytosterol rich diets. Circulation *50:* III-48 (1974).

573 Mellinkoff, S. M.; Machella, T. E., and Reinhold, J. P.: The effect of fat-free diet causing low serum cholesterol. Am. J. med. Sci. *220:* 203–207 (1950).

574 Meltzer, L. E.; Bockman, A. A., and Berryman, G. H.: A means of lowering elevated blood cholesterol levels in patients with previous myocardial infarction. Am. J. med. Sci. *236:* 595–602 (1958).

575 Meneses, H. J.: Dietética de la arterioesclerosis. Med. Rev. Méx. *41:* 73–79 (1961).

576 Menezzi, A. and Morechi, A.: cited (erroneously as Atti acad. Lincei *17:* 95 [1908] and *19:* 126 [1910]) by Bergmann [75].

577 Menschick, W. und Page, I. H.: Über die Resorbierbarkeit des unbestrahlten Ergosterins. Ein Beitrag zur Methodik der quantitativen Ergosterinbestimmung. Hoppe-Seyler's Z. physiol. Chem. *211:* 246–252 (1932).

578 Mercer, E. L.: Double-labelling techniques applied to biosynthetic studies on plant sterol and related compounds. Biochem. J. *96:* 17–18P (1965).

579 Merrill, G. M.: Effects of an unsaturated and saturated lipid on experimental atherosclerosis in rabbits. Proc. Soc. exp. Biol. Med. *105:* 268–270 (1960).

580 Meshcherskaia, K. A. and Borodina, G. P.: Participation of bile acids in the hypocholesterolemic effect of β-sitosterol (Russian). Farm. Toksik. *25:* 44–47 (1962).

581 Meshcherskaia, K. A.; Borodina, G. P.; Koroleva, N. P.; Litvak, F. I., and Astrovskaia, L. A.: Effect of β-sitosterol on the course of experimentally induced atherosclerosis in rats and rabbits (Russian). Farm. Toksik. *22:* 434–440 (1959).

582 Meshcherskaia, K. A.; Borodina, G. P.; Koroleva, N. P., and Sokolova, I. I.: Influence of plant sterols on the course of experimental atherosclerosis in rats and rabbits (Russian). Tr. Vtoroi Nauch. konf. po Vopr. Probl. Zhira a Pitanii, Prom. Leningrad, 237–242 (1962); Chem. Abstr. *6:* 16675 d (1964).

583 Meshcherskaia, K. A.; Koroleva, N. P., and Borodina, G. P.: The effect of lignocerine on the treatment of experimental atherosclerosis in rats (Russian). Farm. Toksik. *24:* 583–586 (1961).

584 Michi, K.; Nakayama, T., and Ezawa, I.: Effect of soybean sterols and corn sterols on the plasma and liver cholesterol levels (Japanese). Eiyǒ Syokurya (Jap. Soc. Food Nutr.) *18:* 436–439 (1966).

585 Miettinen, T.A.: Cholestanol and plant sterols in the adrenal es,gland the rat. Acta chem. scand. *21:* 286–290 (1967).

586 Miettinen, T.A.; Ahrens, E.H., Jr., and Grundy, S.M.: Quantitative isolation and gas-liquid chromatographic analysis of total dietary and fecal neutral steroids. J. Lipid Res. *6:* 411–424 (1965).

587 Migicovsky, B.B.: Control of cholesterol metabolism; in Dock and Snapper, Advances in internal medicine, vol. 11, pp. 137–162 (New York Medical Publishers, New York 1962).

588 Miller, E.C.: General management of atherosclerosis. N. C. med. J. *19:* 149–152 (1958).

589 Miller, E.C.; Mence, H., and Denton, C.A.: The effect of dietary fat and oil on serum and liver cholesterol in female chicks. Fed. Proc. *20:* 367 (1961).

590 Miller, E.C.; Mence, H., and Denton, C.A.: Effect of type dietary fat on plasma and liver cholesterol concentration in female chicks. J. Nutr. *75:* 367–372 (1961).

591 Miller, G.J. and Miller, N.E.: Plasma high density lipoprotein concentration and development of ischemic heart disease. Lancet *i:* 16–19 (1975).

592 Miller, J.P.; Frost, D.V.; Olsen, R.T.; Christianson, B.A., and Lambert, G.F.: Effect of safflower oil and safflower oil plus sitosterol upon cholesterolemia and aortic sudanophilia in rabbits. Fed. Proc. *17:* 485 (1958).

593 Miller, J.P.; Lambert, F., and Frost, D.V.: Regression studies with safflower oil and sitosterol in rabbit atherosclerosis. Circulation Res. *7:* 779–786 (1959).

594 Minato, A. and Otomo, S.: Studies on the constituents of turtle oil. V. Sterol composition of turtle oil (Japanese). Yakugaku Zasshi (J. pharmac. Soc. Jap.) *89:* 1056–1060 (1969).

595 Misra, P.S.; Bhakuni, D.S.; Sharma, V.N., and Kaul, K.N.: Utilization of *Argemone mexicana* in lowering alkalinity of soil. Vigyan Praguti *10:* 215–217 (1961).

596 Morimoto, K.: Stero-bile acids and bile alcohols. LXXXVI. Studies on the sterols in toad liver. Hiroshima, J. med. Sci. *15:* 145–151 (1966).

597 Moses, C.: Pharmacology of drugs used in the control of hypercholesterolemia. Angiology *13:* 59–68 (1962).

598 Mitschek, G.H.A.: Über die Wirkung von *Solanum melongena* bei experimenteller Atheromatose. Q. Jl Crude Drug Res. *12:* 1972–1976 (1972).

599 Mugler, A.: Effet sur l'hypercholestérolémie et l'hyperlipémie du sitostérol à doses moyennes au cours d'une cure hydro-minérale. Strasbourg méd. *13:* 542–546, 548–549, 551–555 (1962).

600 Mühlfelder, G.; Mühlfelder, O. und Kaffarnik, H.; Beta-Sitosterin bei erfolglos vorbehandelten Patienten mit Hypercholesterinämie. Med. Klin. *71:* 775–778 (1976).

601 Murakami, T.; Itokawa, H.; Matsushima, A., and Ikekawa, W.: Studies on the constituents of the root bark of *Aralia elata* (Mia) Seeman (Araliaceae) and on 'β-sitosterol' (Japanese). Yakugaku Zasshi (J. pharmacol. Soc. Japan) *83:* 427–431 (1963).

602 Murphy, E.A.; Rowsell, J.C., and Mustard, J.F.: The effects of sitosterol on serum cholesterol, platelet economy, thrombogenesis and atherosclerosis in the rabbit. Atherosclerosis *17:* 257–268 (1973).

603 Murthy, S.K. and Ganguly, J.: Studies on cholesterol esterase of the small intestine and pancreas of rats. Biochem. J. *83:* 460–469 (1962).

604 Nagy, S. and Nordby, H.E.: Distribution of free and conjugated sterols in orange and tanger juice sacs. Lipids *6:* 826–830 (1971).

605 Nath, N. and Harper, A. E.: Effect of β-sitosterol on serum cholesterol concentration in the rat. Am. J. Physiol. *197:* 102–104 (1959).

606 Nath, N.; Harper, A. E., and Elvehjem, C. A.: Diet and cholesteremia. III. Effect of dietary proteins with particular reference to lipids in wheat gluten. Can. J. Biochem. *37:* 1375–1384 (1959).

607 Nath, N.; Seider, J. C., and Harper, A. E.: Diet and cholesterolemia. VI. Comparative effects of wheat gluten lipids and some other lipids in presense of adequate dietary protein. J. Nutr. *74:* 389–396 (1961).

608 Nath, N.; Wiener, R.; Harper, A. E., and Elvehjem, C. A.: Diet and cholesteremia. I. Development of a diet for the study of nutritional factors affecting cholesteremia in the rat. J. Nutr. *67:* 289–307 (1959).

609 Nes, W. R.; Thampi, N. S., and Lin, J. T.: Phytosterols in normal and tumor-bearing rats. Cancer Res. *32:* 1264–1266 (1972).

610 Nesheim, M. C.; Garlich, J. D., and Hopkins, D. T.: Studies on the effect of soybean meal on fat absorption in young chicks. J. Nutr. *78:* 89–94 (1962).

611 Nicholas, H. J.: Biosynthesis of ^{14}C-sclareol and β-sitosterol from 2-^{14}C-mevalonic acid. Nature, Lond. *189:* 143–144 (1961).

612 Nicol, T. and Bilbey, D. L. J.: Substances depressing the phagocytic activity of the reticulo-endothelial system. Nature, Lond. *182:* 606 (1958).

613 Nicolaysen, R. and Ragard, R.: Effect of various oils and fats on serum cholesterol in experimental hypercholesterolemic rats. J. Nutr. *73:* 299–307 (1961).

614 Nikkari, T.; Schreibman, P. H., and Ahrens, E. H., Jr.: *In vivo* studies of sterol and squalene secretion by the skin. J. Lipid Res. *15:* 563–573 (1974).

615 Nikuni, J.: The physiological action of phytosterol esters (Japanese). J. agric. chem. Soc. Jap. *7:* 827–838 (1932).

616 Nishida, T.; Takenaka, F., and Kummerow, F. A.: Effect of dietary protein and heated fat on serum cholesterol and beta-lipoprotein levels, and on the incidence of experimental atherosclerosis in chicks. Circulation Res. *6:* 194–202 (1956).

617 Nishioka, I.; Ikekawa, N.; Yogi, A.; Kawasaki, T., and Tsukamoto, T.: Plant sterols and triterpenes. II. Separation of stigmasterol, β-sitosterol and campesterol and about so-called 'γ-sitosterol'. Chem. pharm. Bull., Tokyo *13:* 379–384 (1965).

618 Nutrition Reviews: Effect of plant versus animal fat on blood lipids. Nutr. Rev. *13:* 44–45 (1955).

619 Nutrition Reviews: The effect of sitosterols on the serum cholesterol of man. Nutr. Rev. *14:* 39–41 (1956).

620 Nutrition Reviews: Plant sterol absorption in the rat. Nutr. Rev. *14:* 285–286 (1956).

621 Nutrition Reviews: Unsaturated fatty acids and serum cholesterol levels. Nutr. Rev. *14:* 327–328 (1956).

622 Nutrition Reviews: Dietary fat and plasma cholesterol. Nutr. Rev. *16:* 68–70 (1958).

623 Nutrition Reviews: Factors controlling cholesterol excretion. Nutr. Rev. *18:* 19–21 (1960).

624 Nutrition Reviews: Effects of thiouracil and sitosterol on diet-induced hypercholesterolemia and lipomatous arterial lesions in the rat. Nutr. Rev. *18:* 26–27 (1960).

625 Nutrition Reviews: Experimental atherosclerosis in dogs with aortic grafts. Nutr. Rev. *19:* 338–340 (1961).

626 Nutrition Reviews: Effects of animal versus vegetable protein on serum lipids of man. Nutr. Rev. *19:* 70–71 (1961).

627 Nutrition Reviews: The effect of corn oil on fecal sitosterol and cholesterol in man. Nutr. Rev. *22:* 35–38 (1964).

628 Nutrition Reviews: Hypocholesterolemic effect of sitosterol. Nutr. Rev. *22:* 326–328 (1964).

629 Nutrition Reviews: Heated oils and experimental atherosclerosis *26:* 244–245 (1968).

630 Oaks, W.; Lisan, P., and Moyer, J. H.: Management of hypercholesterolemia. Geriatrics *16:* 343–350 (1961).

631 Offhaus, K.: Der Einfluss von wachstumfördernden Faktoren auf die Insektenentwicklung unter besonderer Berücksichtigung der Phyto-Hormone. Z. vergl. Physiol. *27:* 384–428 (1939).

632 Okey, R. and Lyman, M. M.: Dietary fat and cholesterol metabolism. I. Comparative effect of coconut and cottonseed oils at three levels of intake. J. Nutr. *61:* 523–535 (1957).

633 Okey, R.; Lyman, M. M.; Harris, A. G.; Einset, B., and Hain, W.: Dietary fat and cholesterol metabolism: effects of unsaturation of dietary fats on liver and serum lipids. Metabolism *8:* 241–255 (1959).

634 Okey, R. and Stone, M. M.: Lard vs. a vegetable fat in relation to liver and serum cholesterol. J. Am. diet. Ass. *32:* 807–809 (1952).

635 Olson, R. E. and Vester, J. W.: Nutrition-endocrine interrelationships in the control of fat transport in man. Physiol. Rev. *40:* 677–733 (1960).

636 Opdyke, D. F. and Walther, H. O.: Influence of source of cholesterol, grade of cottonseed oil, and breed on experimental atherosclerosis. Proc. Soc. exp. Biol. Med. *85:* 414–415 (1954).

637 Oshima, S. and Suzuki, S.: Influence of several lipids on human serum cholesterol. 2. Effect of safflower oil and butter (Japanese). Eiyogaku Zasshi (Jap. J. Nutr.) *17:* 225–228 (1957).

638 Oster, P.; Schlierf, G.; Heuck, C.; Greten, H.; Gundert-Remy, U.; Haase, W.; Klose, G.; Nothelfer, A.; Raetzer, H., and Schellenberg, B.: Sitosterol in the type II hyperlipoproteinemia; in Greten, Lipoprotein metabolism, pp. 125–130 (Springer, Berlin 1976).

639 Oster, P.; Schlierf, G.; Heuck, C.; Greten, H.; Gundert-Remy, U.; Haase, W.; Klose, G.; Nothelfer, A.; Raetzer, H., and Schellenberg, B.: Sitosterol treatment of familial type II hypercholesterolemia in children and adults; in Schettler, Goto, Hata and Klose, Atherosclerosis, vol. IV, p. 552 (Springer, New York 1977).

640 Page, I. H.: Die Verteilung von Ergosterin in tierischen Geweben nach seiner Verfütterung. Biochem. Z. *220:* 420–431 (1930).

641 Page, I. H.: Treatment of disorders of cholesterol metabolism; in Cook, Cholesterol. Chemistry, biochemistry and pathology, pp. 427–434 (Academic Press, New York 1958).

642 Page, I. H. und Menschick, W.: Über die Wirkung von Ergosterinverfütterung bei Kaninchen. Biochem. Z. *221:* 6–10 (1930).

643 Paquet, C. et Tassel, M.: Sur l'insaponifiable de l'huile d'avocat *(Persea americana).* Oléagineux *21:* 453–454 (1966).

644 Paye, C. M.: Le stigmastérol. Nouvelle vitamine liposoluble. Gaz. méd. Fr. *63:* 673 (1956).

645 Pestel, M.: Indications thérapeutiques de l'hypercholestérolémie primitive. Presse méd. *68:* 653–654 (1960).

646 Peterson, D. W.: Effect of soybean sterols in the diet on plasma and liver cholesterol in chicks. Proc. Soc. exp. Biol. Med. *78:* 143–147 (1951).

647 Peterson, D. W.: Plant sterols and tissue cholesterol levels. Am. J. clin. Nutr. *6:* 644–649 (1958).

648 Peterson, D. W.; Nichols, C. W., Jr.; Peek, N. F., and Chaikoff, I. L.: Depression of plasma cholesterol in human subjects consuming butter containing soy sterols. Fed. Proc. *15:* 569 (1956).

649 Peterson, D. W.; Nichols, C. W., Jr., and Shneour, E. A.: Some relationships among dietary sterols, plasma and liver cholesterol levels, and atherosclerosis in chicks. J. Nutr. *47:* 57–65 (1952).

650 Peterson, D. W.; Shneour, E. A., and Peek, N. F.: Effect of feeding cholesterol esters on plasma and liver cholesterol in the chick. Fed. Proc. *12:* 426 (1953).

651 Peterson, D. W.; Shneour, E. A., and Peek, N. F.: Effect of dietary sterols and sterol esters on plasma and liver cholesterol in the chick. J. Nutr. *53:* 451–459 (1954).

652 Peterson, D. W.; Shneour, E. A.; Peek, N. F., and Gaffey, H. W.: Dietary constituents affecting plasma and liver cholesterol in cholesterol-fed chicks. J. Nutr. *50:* 191–201 (1953).

653 Peterson, J. E. and Hirst, A. E.: Studies on the relation of diet, cholesterol and atheroma in chickens. Circulation *3:* 116–119 (1951).

654 Peterson, J. E.; Wilcox, A. A.; Haley, M. I., and Keith, R. A.: Hourly variation in total serum cholesterol. Circulation *22:* 247–253 (1960).

655 Pezold, F. A.: Über den derzeitigen Stand unserer Kenntnis von der Pathogenese der Atherosklerose im Hinblick auf prophylaktische und kurative Möglichkeiten. Wien. Z. inn. Med. *43:* 137–159 (1962).

656 Physicians' Desk Reference; 31st ed., pp. 582–583, 1048–1049 (Medical Economics, Oradell 1977).

657 Pick, R.; Jain, S.; Kakita, C., and Johnson, P.: Effect of defatted brain extract and soy sterols on plasma cholesterol levels and atherogenesis in cholesterol-oil fed cockerels. Proc. Soc. exp. Biol. Med. *119:* 850–854 (1965).

658 Pick, R.; Katz, L. N.; Rodbard, S., and Stamler, J.: Influences of neutral fat and plant sterols on exogenous and endogenous hypercholesterolemia and hyperphospholipemia in chicks. Circulation *10:* 610 (1954).

659 Editoriale: Stato attuale della terapia ipocolesterolemizzante. Policlinico, Sez. prat. *68:* 1815–1818 (1961).

660 Pollak, O. J.: Prevention of hypercholesterolemia in the rabbit: successful prevention of cholesterol atherosclerosis. Reduction of blood cholesterol in man. Circulation *6:* 459 (1952).

661 Pollak, O. J.: Alcune veduti non ortodosse sull'aterosclerosi. Gaz. sanit. *24:* 1–8 (1953).

662 Pollak, O. J.: Successful prevention of experimental hypercholesteremia and cholesterol atherosclerosis in the rabbit. Circulation *7:* 696–701 (1953).

663 Pollak, O. J.: Reduction of blood cholesterol in man. Circulation *7:* 702–706 (1953).

664 Pollak, O. J.: Combined beta-sitosterol and linoleic acid regimen for cholesterol-fed rabbits. J. Geront. *13:* 140–143 (1958).

665 Pollak, O.J.: Diet and atherosclerosis. Variations on a theme. Am. J. clin. Nutr. 7: 502–507 (1959).

666 Pollak, O.J.: Editorial. Metabolism and function of arterial cells in culture. Paroi Artérielle 2: 145–148 (1974).

667 Pollak, O.J.: Unpublished.

668 Pomeranze, J.: Infant-feeding practices and blood cholesterol levels. Am. J. clin. Nutr. 8: 340–343 (1960).

669 Pomeranze, J. and Chessin, M.: Decholesterolizing agents. Am. Heart J. 49: 262–266 (1955).

670 Portman, O.W.: Dietary factors and the serum cholesterol level. Postgrad. Med. 26: 386–391 (1959).

671 Portman, O.W.; Hegsted, D.M.; Stare, F.J.; Bruno, D.; Murphy, R., and Sinisterra, L.: Effect of the level and type of dietary fat on the metabolism of cholesterol and beta-lipoprotein in the Cebus monkey. J. exp. Med. 104: 817–828 (1956).

672 Portman, O.W. and Stare, F.J.: Dietary regulation of serum cholesterol levels. Physiol. Rev. 39: 407–442 (1959).

673 Pottenger, F.M., Jr. and Krohn, B.: Reduction of hypercholesterolemia by high-fat diet plus soybean phospholipids. Am. J. dig. Dis. 19: 107–109 (1952).

674 Potter, G.C.; Hensley, G.W., and Bruins, H.W.: The effect of oat oil on serum cholesterol in rats fed a hypercholesteremic diet. Fed. Proc. 22: 268 (1963).

675 Prichard, R.W.; Clarkson, T.B.; Lofland, H.B.; Goodman, H.O.; Herndon, C.N., and Netsky, M.G.: Studies on the atherosclerotic pigeon. J. Am. med. Ass. 179: 49–52 (1962).

676 Prista, L.N. and Alves, A.C.: Estuda fitoquímico das folhas de Persea americana Mill. Gard-Dict. García de Orta (Lisbon) 9: 501–503 (1961).

677 Communiqué: Etudes récentes sur le cholestérol, les lipides et l'artériosclérose. A propos du traitement par le sitostérol. Prod. Probl. pharm. 17: LXVIII–LXIX (1962).

678 Quintão, E.; Grundy, S.M., and Ahrens, E.H., Jr.: An evaluation of four methods for measuring cholesterol absorption by the intestine in man. J. Lipid Res. 12: 221–232 (1971).

679 Raicht, R.F.; Cohen, B.I.; Shefer, S., and Mosbach, E.H.: Sterol balance studies in the rat. Effects of dietary cholesterol and β-sitosterol on sterol balance and rate-limiting enzymes of sterol metabolism. Biochim. biophys. Acta 888: 374–384 (1975).

680 Rajalaskshmi, S.; Sarma, D.S.R., and Sarma, P.S.: Effect of rice bran oil on cholesterol metabolism in albino rats. J. sci. industr. Res. 21: C171–173 (1962).

681 Randouin, L. et Brun, P.: Essai d'interprétation de l'influence éventuelle des corps gras alimentaires sur les accidents cardio-vasculaires en fonction du métabolisme intermédiaire. Revue fr. Cps gras 7: 639–653 (1960).

682 Rathman, D.M.: Vegetable oils in nutrition with special reference to unsaturated fatty acids (Corn Products Refining Co., New York 1957).

683 Rees, H.H.; Mercer, E.I., and Goodwin, T.W.: Stereospecific biosynthesis of plant sterols and β-amyrin. Biochem. J. 96: 30–31P (1965).

684 Rees, H.H.; Mercer, E.I., and Goodwin, T.W.: The stereospecific biosynthesis of plant sterols and α- and β-amyrin. Biochem. J. 99: 726–734 (1966).

685 Reeves, J.E.: Hypercholesterolemia: treatment with sitosterol and a low cholesterol diet. Am. Practit. 10: 1193–1197 (1959).

686 Reiner, E.; Topliff, J., and Wood, J. D.: Hypocholesteremic agents derived from sterols of marine algae. Can. J. Biochem. *40:* 1401–1406 (1962).

687 Reiss, F. and Jaimovich, L.: The influence of sitosterol upon psoriasis vulgaris; observations on electrophoretic pattern. Dermatologica *117:* 393–401 (1958).

688 Reitan, R. M. and Shipley, R. E.: The relationship of serum cholesterol to psychologic abilities. J. Geront. *18:* 350–357 (1963).

689 Reitz, R. C. and Hamilton, J. C.: The isolation and identification of two sterols from two species of blue-green algae. Compar. Biochem. Physiol. *25:* 401–406 (1968).

690 Rhoads, D. V. and Barker, N. W.: Effect of dietary fat and additional ingestion of corn oil on hypercholesterolemia. Proc. Mayo Clin. *34:* 225–229 (1959).

691 Rieckmann, P. and Haack, E. (assigned to C. F. Boehringer & Soehne GmbH, Mannheim-Waldhof, Germany): Method of production of therapeutically valuable sitosterol preparations. Germ. 1,131,845 (June 20, 1962).

692 Riley, F. P. and Steiner, A.: Effect of sitosterol on the concentration of serum lipids in patients with coronary atherosclerosis. Circulation *16:* 723–729 (1957).

693 Ristelhueber, J. et Contesse, G.: Etude de l'action sur la surrénale d'un hypocholestérolémiant, la béta-sitostérol. Thérapie *18:* 363–371 (1963).

694 Rivin, A. M.; Yoshino, S.; Shickman, M., and Schjeide, O. E.: Serum cholesterol measurements – hazards in clinical interpretation. J. Am. med. Ass. *166:* 2108–2111 (1958).

695 Robbins, W. E.: Utilization, metabolism and function of sterols in the housefly. Proc. Symp. Radiation Radioisotopes Appl. Insect Agr. Importance, pp. 269–280 (1963).

696 Robbins, W. E.; Dutky, R. C.; Monroe, R. E., and Kaplanis, J. N.: The metabolism of H³-β-sitosterol by the German cockroach. Annls entomol. Soc. Am. *55:* 102–104 (1962).

697 Robbins, W. E.; Shortino, T. J.; Kaplanis, J. N., and Thompson, M. J.: cited by Svoboda et al. (825).

698 Rodbard, S. and Katz, L. N.: Enteric factors in cholesteremia and atherosclerosis. Circulation *8:* 453–454 (1953).

699 Roehm, R. R. and Mayfield, H. L.: Effect of dietary fat on cholesterol metabolism. J. Am. diet. Ass. *40:* 417–421 (1962).

700 Roels, O. A. and Hashim, S. A.: Influence of fatty acids on serum cholesterol. Fed. Proc. *21:* suppl. 11–12, pp. 7–76 (1962).

701 Roffo, A. H.: La berenja como descolesterinizante (*Solanum melongena* L.). Día med. *16:* 798–800, 802–803 (1944).

702 Roffo, A. H.: The egg-plant (*Solanum melongena* L.) in decholesterolization. Yale J. Biol. Med. *18:* 25–30 (1945–1946).

703 Rona, G.; Chappel, C. I., and Gaudry, R.: Aggravation of cholesterol atherosclerosis in rabbits by free unsaturated fatty acids. Can. J. Biochem. *37:* 479–483 (1959).

704 Rosenfeld, R. S. and Hellman, L.: Reduction and esterification of cholesterol and sitosterol by homogenates of feces. J. Lipid Res. *12:* 192–197 (1971).

705 Rosenheim, O. and Webster, T. A.: The metabolism of beta-sitosterol. Biochem. J. *35:* 928–931 (1941).

706 Rosenman, R. H.; Byers, S. O., and Friedman, M.: The effect of soybean sterols on the absorption of cholesterol by the rat. Circulation Res. *2:* 160–163 (1954).

707 Roth, M. et Favarger, P.: La digestibilité des grasses en présence de certains stérols. Helv. physiol. Acta *13:* 249–256 (1955).

708 Rothblat, G. H. and Buchko, M. K.: Effect of exogenous steroids on sterol synthesis in L-cell mouse fibroblasts. J. Lipid. Res. *12:* 647–652 (1971).

709 Rothblat, G. H. and Burns, C. H.: Comparison of the metabolism of cholesterol, cholestanol, and β-sitosterol in L-cell mouse fibroblasts. J. Lipid Res. *12:* 653–661 (1971).

710 Rouser, G.; Nelson, G. J.; Fleisher, S., and Simon, G.: Lipid classes occurring in animal cells; in Chapman, Biologic membranes. Physical fact and function, vol. I, pp. 7–12 (Academic Press, New York 1968).

711 Russell, P. T.; Scott, J. C., and Van Bruggen, J. T.: Effects of dietary fat on cholesterol metabolism in the diabetic rat. J. Nutr. *76:* 460–466 (1962).

712 Sachs, B. A. and Weston, R. E.: Sitosterol administration in normal and hypercholesteremic subjects; the effect in man of sitosterol therapy on serum lipids and lipoproteins. Archs intern. Med. *97:* 738–752 (1956).

713 Said, F.; Said, A. A., and Ibrahim, M. G.: Analytical study of Egyptian oils and fats. VI. Unsaponifiable matter of lettuce seed oil. Egypt. pharm. Bull. *44:* 433–441 (1962 – publ. 1964).

714 Sakurai, K.; Yoshii, E., and Kubo, K.: Gas-liquid chromatography of sterols of Ch'an Su (toad-cake) (Japanese). Yakugaku Zasshi (J. pharmac. Soc. Jap.) *84:* 1166–1171 (1968).

715 Salama, H. S. and El-Sharaby, A. M.: Giberellic acid and β-sitosterol as sterilants of the cotton leaf worm *Spodoptera littoralis* Boisduval. Experientia *28:* 413–414 (1972).

716 Salen, G.; Ahrens, E. H., Jr., and Grundy, S. M.: Quantitative comparisons of absorption and excretion of cholesterol and β-sitosterol in man. J. clin. Invest. *48:* 72–73a (1969).

717 Salen, G.; Ahrens, E. H., Jr., and Grundy, S. M.: Metabolism of β-sitosterol in man. J. clin. Invest. *49:* 952–967 (1970).

718 Sallee, V. L. and Dietschy, J. M.: Determinants of intestinal mucosal uptake of short- and medium-chain fatty acids and alcohols. J. Lipid Res. *14:* 475–484 (1973).

719 Sandquist, M. und Bengtsson, E.: Zur Frage der Bruttoformel des Sitosterins. Ber. dt. chem. Ges. *64:* 2167–2171 (1931).

720 Sandquist, M. und Gorton, J.: Stigmasterin und dessen empirische Formel. Ber. dt. chem. Ges. *63:* 1935–1938 (1930).

721 Schaefer, A. E.; Kowald, J. W.; Wind, S.; Numerof, P.; Kessler, W. B., and McCormack, R. W.: Effect of dietary fat and plant sterol on cholesterol metabolism. Fed. Proc. *14:* 449 (1955).

722 Schaeffer, C. H.; Kaplanis, J. N., and Robbins, W. E.: The relationship of the sterols of Virginia pine sawfly, *Neodiprion pratti* Dyar to those of two host plants, *Pinus virginiana* Mill and *Pinus rigida* Mill. J. insect. Physiol. *11:* 1013–1021 (1965).

723 Schendel, H. E. and Hansen, J. D. L.: Studies on fat metabolism in kwashiorkor. I. Total serum cholesterol. Metabolism *7:* 731–741 (1958).

724 Schettler, G.: Blut- und Organcholesterin der weissen Maus nach Verfütterung pflanzlicher Öle und tierischer Fette mit Phytosterolzusatz. Klin. Wschr. *26:* 566–567 (1948).

725 Schettler, G.: Zur Wirkung der lipotropen Substanzen. Klin. Wschr. *30:* 627–633 (1952).

726 Schettler, G.: Möglichkeiten und Grenzen der Arteriosklerosetherapie. Med. Klin. *57:* 774–780 (1962).

727 Schettler, F.G. and Sanwald, R.: Sitosterol; in Schettler and Boyd, Atherosclerosis. Pathology, physiology, aetiology, diagnosis and clinical management, chap. X, pp. 857–860(–863) (Elsevier, Amsterdam 1969).

728 Schön, H.: Untersuchungen über die Wirkung des Beta-Sitosterins und anderer organischen Substanzen auf den Cholesterinstoffwechsel der Leber. Verh. dt. Ges. inn. Med. *63*: 548–550 (1957).

729 Schön, H.: Neue Untersuchungen über die Wirkungsweise des β-Sitosterins und anderer organischen Verbindungen auf den Cholesterinstoffwechsel. Proc. 4th Congr. Int. Ass. Gerontol., vol. II, pp. 79–81 (1957).

730 Schön, H.: Sterol balance experiments in humans. Nature, Lond. *184*: 1872–1873 (1959).

731 Schön, H.; Demling, L.; Doebl, B.; Bünte, H. und Berg, G.: Untersuchungen über die Wirksamkeit antiatherosklerotischer Präparate. I. Mitteilung: Die Beeinflussung der Serumlipide durch ein Multivitamin-Hormon-Präparat und β-Sitosterin. Med. Klin. *57*: 16–20 (1962).

732 Schön, H. und Engelhardt, P.: Zur Frage der Resorption des β-Sitosterin. Naturwissenschaften *44*: 116 (1957).

733 Schön, H. und Engelhardt, P.: Tierexperimentelle Untersuchungen zur Frage der Resorption von β-Sitosterin. Arzneimittel-Forsch. *10*: 491–496 (1960).

734 Schön, H. und Henning, N.: Untersuchungen zur Regulation des Cholesterinstoffwechsels. Dt. med. Wschr. *84*: 1385–1390 (1959).

735 Schönfeld, L. J. and Sjövall, J.: Identification of bile acids and neutral sterols in guinea pig bile. Bile acids and steroids 163. Acta chem. scand. *20*: 1297–1303 (1966).

736 Schönheimer, R.: Gibt es eine Ablagerung nicht umgewandelter Pflanzensterine im Tierorganismus? Hoppe-Seyler's Z. physiol. Chem. *180*: 16–18 (1929).

737 Schönheimer, R.: Versuch einer Bilanz am Kaninchen bei Verfütterung mit Sitosterin. Hoppe-Seyler's Z. physiol. Chem. *180*: 24–32 (1929).

738 Schönheimer, R.: Über die Sterine des Kaninchenkotes. Hoppe-Seyler's Z. physiol. Chem. *180*: 32–37 (1929).

739 Schönheimer, R.: New contributions in sterol metabolism. Science *74*: 579–584 (1931).

740 Schönheimer, R.: Die Spezifität der Cholesterinresorption und ihre biologische Bedeutung. Klin. Wschr. *11*: 1793–1796 (1932).

741 Schönheimer, R. und Behring, H.v.: Ist unbestrahltes Ergosterin resorbierbar? Klin. Wschr. *9*: 1308 (1930).

742 Schönheimer, R.; Behring, H.v. und Gottberg, K.v.: Ist unbestrahltes Ergosterin resorbiert? Hoppe-Seyler's Z. physiol. Chem. *208*: 77–85 (1932).

743 Schönheimer, R.; Behring, H.v. und Hummel, R.: Über die Spezifität der Resorption von Sterinen, abhängig von ihrer Konstitution. Hoppe-Seyler's Z. physiol. Chem. *192*: 117–124 (1930).

744 Schönheimer, R.; Behring, H.v.; Hummel, R. und Schindel, L.: Über die Bedeutung gesättigter Sterine im Organismus. 4. Mitteilung. Untersuchung der Sterine aus verschiedenen Organen auf ihren Gehalt an gesättigten Sterinen. Hoppe-Seyler's Z. physiol. Chem. *192*: 93–96 (1930).

745 Schönheimer, R. und Dam, H.: Über Ergosterin-Resorption bei der legenden Henne. Hoppe-Seyler's Z. physiol. Chem. *211*: 241–245 (1932).

746 Schönheimer, R. und Yuasa, D.: Speicherungsversuche mit Sitosterin. Hoppe-Seyler's Z. physiol. Chem. *180:* 5–16 (1929).

747 Schönheimer, R. und Yuasa, D.: Verfügt der tierische Organismus über Pflanzensterine? Hoppe-Seyler's Z. physiol. Chem. *180:* 19–23 (1929).

748 Schreiber, K. und Osske, G.: Isolierung von 14α-methyl-5α-stigma-7,24(28)-dien-3β-ol aus *Solanum tuberosum* sowie die Identität dieser Verbindung mit α-Sitosterin. Experientia *19:* 69–70 (1963).

749 Schreiber, K.; Osske, G. und Schembdner, G.: Identifizierung von β-Sitosterin als Hauptsterin des Kartoffelkäfers (*Leptinotarsa decemlineata* Say). Experientia *17:* 463–464 (1961).

750 Schroeder, H.A.: A practical method for the reduction of plasma cholesterol in man. J. chron. Dis. *4:* 461–468 (1956).

751 Schubert, K. und Rose, G.: Über das Vorkommen von β-Sitosterin im Sesamöl und die Anreicherung von Cholesterin im Ölgranuloma. Monatsber. dt. Akad. Wiss., Berlin *6:* 35–37 (1964).

752 Schulman, R.S.; Bhattacharyya, A.K.; Connor, W.E., and Fredrickson, D.S.: β-Sitosterolemia and xanthomatosis. New Engl. J. Med. *294:* 482–483 (1976).

753 Seckfort, H.: Zur therapeutischen Beeinflussung erhöhter Serumlipidwerte. Med. Klin. *57:* 1093–1094, 1097 (1962).

754 Segall, M.M.; Lloyd, J.K.; Fossbrooke, A.S., and Wolff, O.H.: Treatment of familial hypercholesterolemia in children. Lancet *i:* 641–644 (1970).

755 Segall, S. and Neufeld, A.H.: Blood lipid variations in patients with hypercholesteraemia and coronary artery disease. I. Can. med. Ass. J. *83:* 521–524 (1960).

756 Seskind, C.R.; Schroeder, M.T.; Rasmussen, R.A., and Wissler, R.W.: Serum lipid levels in rats fed vegetable oils with and without cholesterol. Proc. Soc. exp. Biol. Med. *100:* 631–634 (1959).

757 Shaper, A.G.: Cardiovascular studies in the Samburu tirbe of North Kenya. Am. Heart J. *63:* 437–442 (1962).

758 Shapiro, W.; Estes, E.H., Jr., and Hilderman, H.L.: The effects of corn oil on serum lipids in normal active subjects. Am. J. Med. *23:* 898–909 (1957).

759 Shefer, S.; Hauser, S.; Lappar, V., and Mosbach, E.H.: Regulatory effect of dietary sterols and bile acids on rat intestinal HMG CoA reductase. J. Lipid Res. *14:* 400–405 (1973).

760 Sherber, D.A.: Hypercholesterolemia. Am. Practit. *8:* 776–783 (1957).

761 Shipley, R.E.: Present status of treatment of atherosclerosis. Proc. Central Section Am. pharm. manuf. Ass. 54–67 (1955).

762 Shipley, R.E.: Symposium on sitosterol. I. The effect of sitosterol ingestion on serum cholesterol concentration. Trans. N.Y. Acad. Sci. *18:* ser. II, pp. 111–118 (1955).

763 Shipley, R.E.: Modern treatment of atherosclerosis. J. Indiana St. med. Ass. *50:* 1623–1628 (1957).

764 Shipley, R.E.; Pfeiffer, R.R.; Marsh, M.M., and Anderson, R.C.: Sitosterol feeding. Chronic animal and clinical toxicology and tissue analysis. Circulation Res. *6:* 373–382 (1958).

765 Shiratari, T. and Goodman, D.S.: Complete hydrolysis of dietary cholesterol esters during intestinal absorption. Biochim. biophys. Acta *106:* 625–627 (1965).

766 Shoppee, C.W.: Chemistry of steroids, 2nd ed., pp. 62–80 (Butterworth, London 1964).

767 Sieber, H.A.; Poe, W.D., and Wheless, J.E.: The problem of treatment of atherosclerosis. Virginia med. Mon. *87:* 442–447 (1960).

768 Siegfried, C.M. and Hyde, P.M.: Comparative metabolism of cholesterol and beta-sitosterol. Fed. Proc. *30:* 1105 (1971).

769 Silber, E.; Pick, R., and Katz, L.N.: Clinical management of atherosclerosis. Circulation *21:* 1193–1204 (1960).

770 Silverman, P.H. and Levinson, Z.H.: Lipide requirements of the larva of *Musca vicina* reared under non-aseptic conditions. Biochem. J. *58:* 291–294 (1954).

771 Simmons, W.J.; Hofman, A.F., and Theodor, E.: Absorption of cholesterol from a micellar solution: intestinal perfusion studies in man. J. clin. Invest. *46:* 874–890 (1967).

772 Sinclair, H.M.: Deficiency of essential fatty acids and atherosclerosis, etcetera. Lancet *270:* 381–383 (1956).

773 Siperstein, M.D.; Chaikoff, I.L., and Reinhardt, W.O.: C^{14}-cholesterol. V. Obligatory function of bile in intestinal absorption of cholesterol. J. biol. Chem. *198:* 111–114 (1952).

774 Sklan, D.; Budowski, P., and Hurwitz, S.: Effect of soy sterols on intestinal absorption and secretion of cholesterol and bile acids in the chick. J. Nutr. *104:* 1086–1090 (1974).

775 Sklarin, B.S.; Seegers, W., and Hirschhorn, K.: Comparative evaluation of nicotinic acid and other forms of therapy in hyperlipemia and hypercholesterolemia. Circulation *24:* 1042–1043 (1961).

776 Smith, P.F.: The role of sterols in the growth and physiology of pleuropneumonia-like organisms. Recent Prog. Microbiol. 518–525 (1963).

777 Smith, P.F.: Relation of sterol structure to utilization in pleuropneumonia-like organisms. J. Lipid Res. *5:* 121–125 (1964).

778 Smith, T.H.F.: Approaches to the treatment of atherosclerosis. Am. J. Pharm. *131:* 369–398 (1959).

779 Šobra, J.: Vrozené vady lipidového metabolismu. VII. Hypocholesterolemický účinek fytosterolů. Čas. lék. čes. *101:* Lék. věda v zahr. *6:* 9–12 (1962).

780 Šobra, J.; Procházka, B.; Sedláčková, E., and Šulc, M.: Vrozené vady lipidového metabolismu. VIII. Familiární hypercholesterolemická xanthomatosa. Vliv fytosterolů na hladinu krevních lipidů. Vnitřní lék. *9:* 642–650 (1963).

781 Sodhi, H.S.; Kudchodkar, B.J.; Varughese, P., and Duncan, D.: Validation of the ratio method for calculating absorption of dietary cholesterol in man. Proc. Soc. exp. Biol. Med. *145:* 107–111 (1974).

782 Soliman, G.: Constituents of *Cynara scolymus.* Egypt. pharm. Bull. *44:* 19–21 (1962).

783 Sperry, W.M. and Bergman, W.: The absorbability of sterols with particular reference to ostreasterol. J. biol. Chem. *119:* 171–176 (1937).

784 Stamler, J.: Basic research on atherosclerosis. Proc. Central Sect. Am. pharm. manuf. Ass. 34–47 (1955).

785 Stamler, J.: Evaluation of current approaches to the therapy of atherosclerotic disease. Trans. 2nd Conf. Cerebral Vasc. Dis., pp.98–124 (Grune & Stratton, New York 1958).

786 Stamler, J.: Diet and atherosclerotic disease. II. Animal experimental research. J. Am. diet. Ass. *34:* 814–818 (1958).

787 Stamler, J.; Pick, R., and Katz, L. N.: Effects of various fats on cholesterolemia and atherogenesis in cholesterol-fed chicken. Fed. Proc. *16:* 123 (1957).

788 Stamler, J.; Pick, R., and Katz, L. N.: Action of casein, egg albumin and corn germ on cholesterolemia and atherogenesis in cockerels. Fed. Proc. *17:* 155 (1958).

789 Stamler, J.; Pick, R., and Katz, L. N.: Saturated and unsaturated fats. Effects on cholesterolemia and atherogenesis in chicks on high-cholesterol diet. Circulation Res. *7:* 398–402 (1959).

790 Stamnes, P. L.; Palumbo, P. J., and Kottke, B. A.: Serum lipid levels and cholesterol balance in humans: effect of sucrose-starch exchange. Circulation *48:* IV–254 (1973).

791 Stare, F. J.: in Long-term management of patients with coronary artery disease. Clinical conference. Circulation *17:* 945–952 (1958).

792 Stare, F. J.; Van Itallie, T. B.; McCann, M. B., and Portman, O. W.: Nutritional studies relating to serum lipids and atherosclerosis. Therapeutic implications. J. Am. med. Ass. *164:* 1920–1925 (1957).

793 St. Clair, R. W.; Lehner, N. D. M., and Hamm, T. E.: Use of fecal flow markers for the determination of sterol ring degradation in three species of nonhuman primates. Circulation *48:* IV–253 (1973).

794 Steinberg, D.: Chemotherapeutic control of serum lipid levels. Trans. N.Y. Acad. Sci. *24:* 704–723 (1962).

795 Steiner, A. and Dayton, S.: Production of hyperlipemia and early atherosclerosis in rabbits by a high vegetable fat diet. Circulation Res. *4:* 62–66 (1956).

796 Steiner, A. and Riley, F. P.: The effect of beta and dihydro-beta-sitosterol on the serum lipids of patients with coronary atherosclerosis. Circulation *12:* 483 (1955).

797 Steiner, A.; Varson, A., and Ruiman, D.: Effect of a formula diet containing various vegetable oils upon the serum lipids of human subjects. Circulation *16:* 495 (1957).

798 Steiner, A.; Varson, A., and Samuel, P.: Effect of saturated and unsaturated fats on the concentration of serum cholesterol and experimental atherosclerosis. Circulation Res. *7:* 448–453 (1959).

799 Steiner, C. S. and Fritz, E.: Pharmaceutical grade sterols from tall oil. J. Am. Oil Chem. Soc. *36:* 354–357 (1959).

800 Stern, M. (assigned to Eastman Kodak Co., Rochester, N.Y.): Watersoluble phytosterol derivatives (and: Correction of patent). US 3,004,043 (Oct. 10, 1961).

801 Subbiah, M. T. R.: The metabolism of plant sterols in the rat; thesis, Toronto (1971).

802 Subbiah, M. T. R.: Subject review. Significance of dietary plant sterols in man and experimental animals. Proc. Mayo Clin. *46:* 549–559 (1971).

803 Subbiah, M. T. R.: Analysis of sterols: application to sterol balance studies. Am. J. clin. Nutr. *25:* 780–788 (1972).

804 Subbiah, M. T. R.: Dietary plant sterols: current status in human and animal sterol metabolism. Am. J. clin. Nutr. *26:* 219–225 (1973).

805 Subbiah, M. T. R.: Cholestanol and plant sterols in pigeon skin. Lipids *8:* 158–160 (1973).

806 Subbiah, M. T. R. and Kottke, B. A.: Biliary and fecal steroids of atherosclerosis susceptible and atherosclerosis resistant pigeons. Circulation *42:* III–25 (1970).

807 Subbiah, M. T. R. and Kottke, B. A.: Comparison of sterol composition and transformation in chicken and pigeon intestines. Experientia *28:* 1431–1432 (1972).

808 Subbiah, M. T. R.; Kottke, B. A., and Carlo, I. A.: Experimental studies in the spontaneous-atherosclerosis-susceptible white Carneau pigeon; nature of biliary and fecal neutral steroids. Proc. Mayo Clin. *45:* 729–737 (1970).

809 Subbiah, M. T. R.; Kottke, B. A., and Carlo, I. A.: Uptake of campesterol in pigeon intestine. Biochim. biophys. Acta *249:* 643–646 (1971).

810 Subbiah, M. T. R.; Kottke, B. A., and Zoliman, P. E.: Fecal sterols of some avian species. Compar. Biochem. Physiol. *41B:* 695–704 (1972).

811 Subbiah, M. T. R. and Kuksis, A.: Fate of intravenously administered β-sitosterol-22-23-H³ in the rat. Proc. Can. Fed. biol. Soc. *11:* 140 (1968).

812 Subbiah, M. T. R. and Kuksis, A.: Oxidation of 4-¹⁴C-beta-sitosterol by mitochondria of rat liver and testes. Fed. Proc. *28:* 515 (1969).

813 Subbiah, M. T. R. and Kuksis, A.: Metabolism of β-sitosterol-4-C¹⁴ in the rat liver. Proc. Can. Fed. biol. Soc. *12:* 69 (1969).

814 Subbiah, M. T. R. and Kuksis, A.: Differences in metabolism of cholesterol and sitosterol following intravenous injection in rats. Biochim. biophys. Acta *306:* 96–105 (1973).

815 Subbiah, M. T. R.; Naylor, M. C., and Kottke, B.: Intestinal alteration of sterols in patients: implications for balance studies; in Schettler and Weigel, Proc. 3rd Int. Symp. Atheroscl., pp. 295–296 (Springer, New York 1974).

816 Sugano, M.; Ikeda, I., and Marioka, H.: A comparison of hypocholesterolemic activity of β-sitosterol and β-sitostanol in rats; in Schettler, Goto, Hata and Klose, Atherosclerosis, vol. IV, p. 292 (Springer, New York 1977).

817 Sugano, M.; Kamo, F.; Ikeda, I., and Morioka, H.: Lipid-lowering activity of phytosterols in rats. Atherosclerosis *24:* 301–309 (1976).

818 Supniewski, J.: Leczenie miażdżycy. Polski tydog. Lekar. *15:* 844–851 (1960).

919 Suzuki, S.; Oshima, S., and Yamakawa, K.: Influence of several lipids on human serum cholesterol. III. Effect of several vegetable oils (Japanese). Eiyogaku Zasshi (Jap. J. Nutr.) *19:* 230–231 (1961).

820 Suzuki, S.; Oshima, S.; Yamakawa, K.; Kuga, T., and Terada, K.: Influence of several lipids on human cholesterol. IV. Effect of several margarines on the market (Japanese). Eiyogaku Zasshi (Jap. J. Nutr.) *19:* 232–234 (1961).

821 Suzuki, S.; Tezuka, T.; Kajiwara, S.; Kuga, T., and Mitani, M.: Influence of several lipids on human serum cholesterol. V. Effect of rice oil (Japanese). Eiyogaku Zasshi (Jap. J. Nutr.) *20:* 139–141 (1962).

822 Suzuki, S.; Tezuka, T.; Kajiwara, S.; Kuga, T., and Mitani, M.: Influence of several lipids on human serum cholesterol. VI. Effect of several margarines on trial (Japanese). Eiyogaku Zasshi (Jap. J. Nutr.) *20:* 166–168 (1962).

823 Suzuki, S.; Tezuka, T.; Kajiwara, S.; Kuga, T., and Mitani, M.: Influence of several lipids on human serum cholesterol. VII. Effect of several margarines on trial 2 (Japanese). Eiyogaku Zasshi (Jap. J. Nutr.) *20:* 172–174 (1962).

824 Svoboda, J. A.; Hutchins, R. F. N.; Thompson, M. J., and Robbins, W. E.: 22-trans-cholesta-5,22,24-trien-3β-ol – an intermediate in the conversion of stigmasterol to cholesterol in the tobacco hornworm, *Manduca sexta* (Johannson). Steroids *14:* 469–476 (1969).

825 Svoboda, J. A. and Robbins, W. E.: Conversion of beta-sitosterol to cholesterol blocked in an insect by hypocholesterolemic agents. Science *156:* 1637–1638 (1967).

826 Svoboda, J. A. and Robbins, W. E.: Desmosterol as a common intermediate in the

conversion of a number of C_{28} and C_{29} plant sterols to cholesterol by the tobacco hornworm. Experientia 24: 1131–1132 (1968).

827 Svoboda, J.A. and Robbins, W.E.: The inhibitive effect of azasterols on sterol metabolism and development in insects with special reference to tobacco hornworm. Lipids 6: 113–119 (1971).

828 Svoboda, J.A.; Robbins, W.E.; Cohen, C.F., and Shortino, T.J.: Phytosterol utilization and metabolism in insects: recent studies with *Tribolium confusum*. Insect Mite Nutr. 1972: 505–516.

829 Svoboda, J.A.; Thompson, M.J.; Elden, T.C., and Robbins, W.E.: Unusual composition of sterols in a phytophagous insect, the Mexican bean beetle reared on soybean plants. Lipids 9: 752–755 (1974).

830 Svoboda, J.A.; Thompson, M.J., and Robbins, W.E.: Desmosterol, an intermediate in dealkylation of β-sitosterol in the tobacco hornworm. Life Sci. 6: 395–404 (1967).

831 Svoboda, J.A.; Thompson, M.J., and Robbins, W.E.: 3β-hydroxy-24-norchol-5-en-23oic acid – a new inhibitor of the Δ^{24}-sterol reductase enzyme system(s) in the tobacco hornworm, *Manduca sexta* (Johannson). Steroids 12: 559–570 (1968).

832 Svoboda, J.A.; Thompson, M.J., and Robbins, W.E.: Azasteroids: potent inhibitors of insect molting and metamorphosis. Lipids 7: 553–556 (1972).

833 Svoboda, J.A.; Womack, M.; Thompson, M.J., and Robbins, W.E.: Comparative studies on the activity of 3β-hydroxy-Δ^5-norcholenic acid on the Δ^{24}-sterol enzyme(s) in an insect and in the rat. Compar. Biochem. Physiol. 30: 541–549 (1969).

834 Swell, L.; Boiter, T.A.; Field, H., Jr., and Treadwell, C.R.: Esterification of soybean sterols *in vitro* and their influence on blood cholesterol level. Proc. Soc. exp. Biol. Med. 86: 295–298 (1954).

835 Swell, L.; Boiter, T.A.; Field, H., Jr., and Treadwell, C.R.: The absorption of plant sterols and their effect on serum and liver sterol levels. J. Nutr. 58: 385–398 (1956).

836 Swell, L.; Byron, J.E., and Treadwell, C.R.: Cholesterol esterase. IV. Cholesterol-esterase of rat intestinal mucosa. J. biol. Chem. 186: 543–548 (1950).

837 Swell, L.; Field, H., Jr., and Treadwell, C.R.: Sterol specificity of pancreatic cholesterol esterase. Proc. Soc. exp. Biol. Med. 87: 216–218 (1954).

838 Swell, L.; Flick, D.F.; Field, H., Jr., and Treadwell, C.R.: Role of fat and fatty acids in absorption of dietary cholesterol. Am. J. Physiol. 180: 124–128 (1955).

839 Swell, L.; Schools, P.E., Jr., and Treadwell, C.R.: Influence of a family diet pattern high in linoleic acid on serum cholesterol level: one-year study. Proc. Soc. exp. Biol. Med. 111: 48–50 (1962).

840 Swell, L.; Schools, P.E., Jr., and Treadwell, C.R.: Family diet pattern lowering the serum cholesterol level. Am. J. clin. Nutr. 11: 102–107 (1962).

841 Swell, L.; Stutzman, E.; Law, M.D., and Treadwell, C.R.: Intestinal absorption of cholesterol-4-C^{14}-*D*-glucoside. Archs Biochem. 97: 383–386 (1962).

842 Swell, L. and Treadwell, C.R.: Metabolic fate of injected C^{14}-phytosterols. Proc. Soc. exp. Biol. Med. 108: 810–813 (1961).

843 Swell, L. and Treadwell, C.R.: Cholesterol esterases. II. Characterization of the esterifying cholesterol esterase of pancreatin. J. biol. Chem. 182: 479–487 (1950).

844 Swell, L. and Treadwell, C.R.: Cholesterol esterases. VI. Relative specificity of pancreatic cholesterol esterase. J. biol. Chem. 212: 141–150 (1955).

845 Swell, L.; Trout, E.C., Jr.; Field, H., Jr., and Treadwell, C.R.: Absorption of H^3-β-sitosterol in the lymph fistula rat. Proc. Soc. exp. Biol. Med. 100: 140–142 (1959).

846   Swell, L.; Trout, E.C., Jr.; Field, H., Jr., and Treadwell, C.R.: Intestinal metabolism of
      C^{14}-phytosterols. J. biol. Chem. *234:* 2286–2289 (1959).

847   Swell, L.; Trout, E.C., Jr.; Hopper, R., and Field, H., Jr.: The mechanism of cholesterol
      absorption. The influence of hormones on lipid metabolism in relation to arterioscle-
      rosis. Ann. N.Y. Acad. Sci. *72:* 813–825 (1959).

848   Swell, L.; Trout, E.C., Jr.; Vahouny, G.V.; Field, H., Jr.; Schuching, S.v., and
      Treadwell, C.R.: Influence of H^3-β-sitosterol on sterol excretion. Proc. Soc. exp. Biol.
      Med. *97:* 337–339 (1958).

849   Sylvén, C.: Influence of blood supply on lipid uptake from micellar solutions by the rat
      small intestine. Biochim. biophys. Acta *203:* 365–375 (1970).

850   Sylvén, C.: Uptake of micellar lipids by small-intestinal segments (under different
      experimental conditions). Acta physiol. scand. *83:* 289–299 (1971).

851   Sylvén, C. and Borgström, B.: Absorption and lymphatic transport of cholesterol and
      sitosterol in the rat. J. Lipid Res. *10:* 179–182 (1969).

852   Sylvén, C. and Nordström, C.: The site of absorption of cholesterol and sitosterol in the
      rat small intestine. Scand. J. Gastroent. *5:* 57–63 (1970).

853   Tamura, T.; Truscott, B., and Idler, D.R.: Sterol metabolism in the oyster. J. Fish. Res.
      Board Can. *21:* 1519–1522 (1964).

854   Tanret, C.: Sur un nouveau principe imméditate de l'ergot de seigle, l'ergostérine. C. r.
      Séanc. Soc. Biol. *108:* 98–100 (1889).

855   Teshima, S.-I.: Bioconversion of β-sitosterol and 24-methylcholesterol to cholesterol in
      marine crustacea. Compar. Biochem. Physiol. *39B:* 815–822 (1971).

856   Teshima, S. and Kanazawa, A.: Bioconversion of the dietary ergosterol to cholesterol
      by *Artemia salina.* Compar. Biochem. Physiol. *38:* 603–607 (1971).

857   Tezuka, T.; Oshima, S., and Suzuki, S.: Influence of several lipids on the serum
      cholesterol and phospholipid level. 1. Effect of corn oil and butter (Japanese). Eiyogaku
      Zasshi (Jap. J. Nutr.) *16:* 183–185 (1958).

858   Thakkar, A.L. and Diller, E.R. (assigned to Eli Lilly & Co, Indianapolis, Ind.): Phar-
      maceutical dispersible powder of sitosterols and a method for the preparation thereof.
      US 3,881,005 (Apr. 29, 1975).

859   Thiers, H.: De l'intérêt biologique et thérapeutique du stigmastérol. Presse méd. *63:*
      820–821 (1955).

860   Thiers, H.: Intérêt thérapeutique des phytostérols. Presse méd. *65:* 1814–1815
      (1957).

861   Thiers, H.; Colomb, D. et Fayolle, J.: Le traitement des affections ostéoarticulaires par
      les phytostéroles et par les insaponifiables d'huiles végétables en solution alcoholique.
      Lyon méd. *91:* 331–338 (1959).

862   Thiers, H.; Zwingelstein, Jouanneteau, Fayolle e Moulin: Un gruppo d'agenti tera-
      peutici nuovi: gli insaponificabili degli olii vegetali, i loro carotenoidi, i loro fitosteroli e
      le loro frazioni indeterminate somministrati in soluzione alcoholica. Minerva med.,
      Roma *52:* 2144–2152 (1961).

863   Thomas, W.A.; Konikov, N.; O'Neal, R., and Lee, K.T.: Saturated versus unsaturated
      fats in experimental arteriosclerosis. Archs Path. *63:* 571–575 (1957).

864   Thompson, M.J.; Kaplanis, J.N.; Robbins, W.E., and Svoboda, J.A.: Metabolism of
      steroids in insects; in Paoletti and Kritchevsky, Adv. Lipid Res., vol.11, pp.219–265
      (Academic Press, New York 1973).

865   Thompson, M.J.; Louloudes, S.J.; Robbins, W.E.; Waters, J.A.; Steele, J.A., and

Mosettig, E.: Identity of the 'house fly sterol'. Biochem. biophys. Res. Commun. *9:* 113–119 (1962).

866 Thompson, M.J.; Robbins, W.E., and Baker, G.L.: The nonhomogeneity of soybean sterols – 'gamma-sitosterol'. Steroids *2:* 505–512 (1963).

867 Thoms, H.: Über Phytosterine. Arch. Pharm., Weinheim *235:* 39–43 (1897).

868 Tidwell, H.C. and Gifford, P.: Effect of ingestion of isomeric fatty acids on cholesterol and lipids of serum and liver. Fed. Proc. *24:* 192 (1965).

869 Tieri, O. e Tocco, G.: Azione inhibitrice dell'isocolesterolo sull'assorbimento del colesterolo. Boll. Soc. ital. Biol. sper. *34:* 387–389 (1958).

870 Tixier, L. et Eck, M.: Action du cynara sur le métabolism du cholestérol. J. méd. Fr. *24:* 56–61 (1935).

871 Tixier, L. et Eck, M.: L'actions physiologiques et thérapeutiques du *'Cynara scolimus'*. Presse méd. *47:* 880–883 (1939).

872 Tobian, L. and Tuna, M.: The efficacy of corn oil in lowering the serum cholesterol of patients with coronary atherosclerosis. Clin. Res. Proc. *5:* 182 (1957).

873 Tobian, L. and Tuna, M.: The efficacy of corn oil in lowering the serum cholesterol of patients with coronary atherosclerosis. Am. J. med. Sci. *235:* 133–137 (1958).

874 Tognoli Lena, E.: Influenza dei fitosteroli sulle lesioni indotte della dieta ipercolesterolica in alcuni organi del coniglio. Boll. chim. farm. *98:* 267–272 (1959).

875 Tolckmitt, W.: Beziehungen zwischen Ernährung und Gefässkrankheiten vom Standpunkt des Tierexperiments. Medizinische 1288–1292, 1412–1416 (1959).

876 Tomkins, G.M.; Sheppard, H., and Chaikoff, I.L.: Cholesterol synthesis by liver. IV. Suppression by steroid administration. J. biol. Chem. *203:* 781–786 (1953).

877 Transbøl, I.; Winkler, K., and Tygdtrup, N.: Triparanolbehandling ved hyperkolesterolaemie. Ugeskr. Laeger *123:* 1167–1169 (1961).

878 Trčka, V.: Výsledky farmakologického hodnocení směsí fytosterolů. Dokumentace VÚFB, Praha (1958). Intramural report not available for distribution; cited by Šobra [779].

879 Tsuda, H.; Akagi, S., and Kishida, Y.: Discovery of cholesterol in some red algae. Science *126:* 927–928 (1957).

880 Turpeinen, O. and Jokipii, S.G.: Effects on serum-cholesterol level of changes in dietary fat composition and of administration of vitamins, thyroid and other substances. J. Atheroscler. Res. *1:* 307–316 (1961).

881 Turpeinen, O.; Miettinen, M.; Karvonen, M.J.; Roine, P.; Pekkarinen, M.; Lehtusuo, E. J. and Altivirta, P.: Dietary prevention of coronary heart disease: long-term experiments. I. Observation on male subjects. Am. J. clin Nutr. *21:* 255–276 (1968).

882 Turpeinen, O.; Roine, P.; Karvonen, M.J.; Runeberg, J.; Pekkarinen, M.; Rautanen, Y., and Altivirta, P.: Effect on serum-cholesterol level of replacement of dietary milk fat by soybean oil. Lancet *i:* 196–198 (1960).

883 Tygstrup, N.; Winkler, K.; Jørgensen, K., and Andersen, B.: Sitosterolbehandling ved hypercholesterolaemie. Ugeskr. Laeger *119:* 1193–1195 (1957).

884 Uzan, A.: Les stérols des sous-produits de l'huilerie: obtention, identification, possibilités d'emploi. Annls Nutr. Aliment. *13:* A, pp.389–416 (1959).

885 Vahouny, G.V. and Treadwell, C.R.: Absorption of cholesterol esters in the lymph-fistula rat. Am. J. Physiol. *195:* 516–520 (1958).

886 Vahouny, G.V. and Treadwell, C.R.: Enzymatic synthesis and hydrolysis of choles-

terol esters; in Glick, Methods of biochemical analysis, vol. 16, pp. 219–272 (Interscience/Wiley & Sons, London 1968).

887 Vaĭsman, N. M. and Georgievskaya, L. M.: Effect of β-sitosterol on the blood lipid levels in patients with coronary atherosclerosis (Russian). Ter. Arkh. *33:* 29–36 (1961).

888 Vaĭsmann, M.; Georgievskaya, L. M., and Khabarova, E. I.: β-Sitosterol, its influence on the lipoprotein fractions of the blood in coronary atherosclerosis (Russian). Ter. Arkh. *35:* 64–68 (1963).

889 Valkema, A. J.: The influence of phytosterols on the absorption of cholesterol. Acta physiol. pharmac. neerl. *4:* 291–292 (1955–1956).

890 Valkema, A. J.: The management of hypercholesterolemia in animals and man. Acta physiol. pharmac. neerl. *4:* 424–425 (1955–1956).

891 Valkema, A. F.: Der Einfluss von Phytosterin auf die Cholesterinresorption. Fette Seifen *59:* 564 (1957).

892 Van Gierke, E.: Discussion of papers I–XIII, Verh. dt. path. Ges. *20:* 159–160 (1925).

893 Van Handel, E.; Neumann, H., and Bloehm, T.: A diet restricted in refined cereals and saturated fats: its effect on the serum-lipid level of atherosclerotic patients. Lancet *272:* 245–246 (1957).

894 Van Itallie, T. B.: Nutritional research in atherosclerosis – a progress report. J. Am. med. Ass. *34:* 248–253 (1958).

895 Van Wagtendonk, W. J. and Conner, R. L.: Steroid requirements for *Paramecium aurelia,* var. 4, stock 51.7(s) in axenic culture. Fed. Proc. *12:* 283 (1953).

896 Van Wagtendonk, W. J.; Conner, R. L.; Miller, C. A., and Rao, M. R. R.: Growth requirements of *Paramecium aurelia,* var. 4, stock 51.7 sensitiveness and killers in axenic medium. Ann. N.Y. Acad. Sci. *56:* 929–937 (1953).

897 Vanzetti, G.; Coronelli, M. e Sfondrini, G.: Ricerche sperimentali sull'influenza degli steroli vegetali e di un agente colagogo sul ricambo del colesterolo. Folia cardiol. *16:* 221–230 (1957).

898 Vecchi, G. P.; Lapiccirella, R.; Guarienti, F. e Saetti, G. C.: Introduzione alla studio di alcuni fattori dietetici e farmacologici utili nella regolazione della colesterolemia. G. Clin. med. *39:* 707–726 (1958).

899 Verdonk, G.: Athéromatose et diététique. Z. ErnährWiss. *2:* 1–23 (1961).

900 Verdonk, G.: Nutrition et athéromatose. Sem. Hôp. Paris *38:* 853–858 (1962).

901 Vesselinovitch, D.; Getz, G. S.; Hughes, E. H., and Wissler, R. W.: Atherosclerosis in the rhesus monkey fed three food fats. Atherosclerosis *20:* 303–321 (1974).

902 Walker, G. R.; Morse, E. H., and Overley, V. A.: The effect of animal protein and vegetable protein diets having the same fat content on the serum lipid levels of young women. J. Nutr. *72:* 317–321 (1960).

903 Wallis, E. S. and Chakravorty, P. N.: The nature of sterols in cottonseed oil. J. org. Chem. *2:* 335–340 (1937).

904 Warembourg, H.; Desruelles, J.; Decalf, A. et Buyck, J.: Essais du beta-sitostérol dans le traitement au long course de l'athérosclérose. Lille méd. *8:* 484–487 (1963).

905 Waters, J. A. and Johnson, D. F.: Biosynthesis of sterols in the soybean plant. Archs Biochem. *112:* 387–391 (1965).

906 Weigand, A. J. (assigned to Intellectual Property Development Corp.): Arzneipräparate mit Cholesterinspiegel-senkender Wirkung. GP 2261571 (USSN 208887 & 304295) (Dec. 16, 1971); P 22 61 571.0 (Dec. 15, 1972; June 28, 1973).

907 Weiner, N.; Walker, W.J., and Milch, L.J.: Effect of beta-sitosterol on serum cholesterol and lipoprotein levels of patients ingesting a low-fat diet. PB 135059, 6 pp. (May,
1958).

908 Weiner, N.; Walker, W.J., and Milch, L.J.: Effect of beta-sitosterol on serum cholesterol and lipoprotein levels of patients ingesting a low-fat diet. Am. J. med. Sci. *235:*
405–409 (1958).

909 Weis, H. and Dietschy, J.M.: The interaction of various control mechanisms in determining the rate of hepatic cholesterogenesis in the rat. Biochim. biophys. Acta *398:*
315–324 (1975).

910 Weiss, J.P.; Johnson, R.M., and Naber, E.C.: Effect of some dietary factors and drugs
on cholesterol concentration in the egg and plasma of the hen. J. Nutr. *91:* 119–128
(1967).

911 Weizel, A.: Beeinflussung des Cholesterinstoffwechsels durch Beta-sitosterin. Med.
Klin. *70:* 242–246 (1975).

912 Wells, A.F. and Alfin-Slater, R.B.: Effect of bile acids and beta-sitosterol on cholesterol
metabolism in essential fatty acid deficient rats. Fed. Proc. *17:* 333 (1958).

913 Wells, V.M. and Bronte-Stewart, B.: Egg yolk and serum-cholesterol levels: importance of dietary cholesterol intake. Br. med. J. *i:* 577–581 (1963).

914 Werbin, H.; Chaikoff, I.L., and Imada, M.R.: Radiochemical purity of β-sitosterol
titrated by catalytic exchange and by the Wilzbach procedure. Archs Biochem. *89:*
213–217 (1960).

915 Werbin, H.; Chaikoff, I.L., and Jones, E.E.: The isolation of H³-cortisol from the urine
of guinea pigs fed H³-labeled β-sitosterol. Fed. Proc. *18:* 350 (1959).

916 Werbin, H.; Chaikoff, I.L., and Jones, E.E.: Isolation of H³-cortisol from the urine of
guinea pigs fed H³-cholesterol. J. biol. Chem. *234:* 282–284 (1959).

917 Werbin, H.; Chaikoff, I.L., and Jones, E.E.: The metabolism of H³-β-sitosterol in the
guinea pig: its conversion to urinary cortisol. J. biol. Chem. *235:* 1629–1633 (1960).

918 West, R.O. and Hayes, O.B.: Diet and serum cholesterol levels. A comparison between
vegetarians and non-vegetarians in a Seventh-day Adventist group. Am. J. clin. Nutr.
21: 853–862 (1963).

919 Whitney, J.O.; Horn, H., and Gordan, G.S.: Quantitation of osteolytic phytosterol
acetate in human serum. Proc. Soc. exp. Biol. Med. *143:* 1000–1002 (1973).

920 Wientjens, W.H.J.M.; Van der Marel, T., and Bennett, E.P.: The influence of several
sterols on the conversion of β-sitosterol into cholesterol in the cockroach. Experientia
27: 373–375 (1971).

921 Wilcox, E.B. and Galloway, L.S.: Serum and liver cholesterol, total lipids and lipid
phosphorus levels of rats under various dietary regimens. Am. J. clin. Nutr. *9:* 236–253
(1961).

922 Wilcox, E.B. and Galloway, L.S.: Serum cholesterol and different dietary fats. J. Am.
diet Ass. *38:* 227–230 (1961).

923 Wilkens, J.A.: The effect of dietary lipid fractions on the serum cholesterol of the rat.
S.Afr. med. J. *32:* 85 (1958).

924 Wilkens, J.A.: A proposed mechanism for the serum cholesterol regulating effect of
dietary fats and oils. S. Afr. med. J. *33:* 1076 (1959).

925 Wilkens, J.A. and Wit, H. de: The effect of dietary lipids on the serum cholesterol of
rats. Can. J. Biochem. *40:* 1079–1090 (1962).

926 Wilkens, J.A.; Wit, H. de, and Bronte-Stewart, B.: A proposed mechanism for the effect

of different dietary fats on some aspects of cholesterol metabolism. Can. J. Biochem. *40:* 1091–1100 (1962).

927 Wilkinson, C. F., Jr.: Effect of lipotropes and sitosterol on the level of blood lipids and the clinical course of angina pectoris. J. Am. Geriat. Soc. *3:* 381–388 (1955).

928 Wilkinson, C. F., Jr.: Drugs other than anticoagulants in treatment of atherosclerotic heart disease. J. Am. med. Ass. *163:* 927–930 (1957).

929 Wilkinson, C. F., Jr.; Boyle, E.; Jackson, R. S., and Benjamin, M. R.: The effect of varying the intake of dietary fat and the ingestion of sitosterol on the lipid fractions of human serum. Circulation *8:* 444 (1953).

930 Wilksinson, C. F., Jr.; Boyle, E.; Jackson, R. S., and Benjamin, M. R.: Effect of varying the intake of dietary fat and the ingestion of sitosterol on lipid and lipoprotein fractions of human serum. Metabolism *4:* 302–309 (1955).

931 Wilkinson, C. F., Jr.; Epstein, F. H.; Keys, A.; Kinsell, L. W.; Pollak, O. J., and Stare, F. J.: Panel discussion on lipid metabolism in cardiovascular disease. J. Am. Geriat. Soc. *6:* 451–470 (1958).

932 Wilkinson, C. F., Jr.; Jackson, R. S.; Bozian, R. C.; Benjamin, M. R.; Levere, A. H.; Craft, G., and Davidson, N. W.: Symposium on sitosterol. II. Clinical experience with 'sitosterols'. Trans. N. Y. Acad. Sci. *18:* ser. II, pp. 119–122 (1955).

933 Wilkinson, C. F., Jr.; Jackson, R. S., and Vogel, W. C.: Effect of feeding dried egg plant *(Solanum melongena* L) on plasma cholesterol. Proc. Soc. exp. Biol. Med. *71:* 656–658 (1949).

934 Williams, G. E. O. and Thomas, G.: Sunflower-seed oil and serum lipids. Lancet *272:* 428–429 (1957).

935 Windaus, A.: cited by Schönheimer et al. [744].

936 Windaus, A. und Hauth, A.: Über Stigmasterin, ein neues Pflanzensterin aus Calabar-Bohnen. Ber. dt. chem. Ges. *39:* 4378–4384 (1906).

937 Wissler, R. W.; Frazier, L. E.; Hughes, R. H., and Rasmussen, R. A.: Atherogenesis in the cebus monkey. I. A comparison of three food fats under controlled dietary conditions. Archs Path. *74:* 312–322 (1962).

938 Wissler, R. W.; Hughes, R. H.; Frazier, L. E., and Rasmussen, R. A.: The effects of feeding fats with varying fatty acid composition on the blood and tissue lipids of Cebus monkey. Circulation *22:* 833 (1960).

939 Wolff, R. et Brignon, J. J.: Recherches éxperimentales sur la résorption intestinale du cholestérol chez le lapin. Bull. Soc. Chim. biol. *38:* 99–109 (1956).

940 Wolff, R.; Herbeuval, R.; Cuny, G.; Brignon, J. J.; Gilgenkrantz, J. et Ulrich, P.: Recherches expérimentales et cliniques au sujet de l'action de l'huile de soja sur la cholestérolémie. Presse méd. *66:* 1706–1708 (1958).

941 Wood, J. D. and Biely, J.: The effect of dietary marine fish oils on the serum cholesterol levels in hypercholesterolemic chickens. Can. J. Biochem. *38:* 19–24 (1960).

942 Wood, P. D. S.; Shioda, R., and Kinsell, L. W.: Dietary regulation of cholesterol metabolism. Lancet *ii:* 604–607 (1966).

943 Wright, L. D.: The anomalous solubility of cholesterol in oils. Proc. Soc exp. Biol. Med. *121:* 265–267 (1966).

944 Wright, L. D.: Influence of a plant sterol preparation on the solubility of cholesterol in triglycerides. Proc. Soc. exp. Biol. Med. *123:* 447–450 (1966).

945 Wright, L. D.: The effect of various compounds on the adsorption of cholesterol in an *in vitro* system. Proc. Soc. exp. Biol. Med. *402:* 405 (1972).

946 Wright, L.D. and Presberg, J.A.: The effect of certain compounds on the solubility of cholesterol in coconut oil. Fed. Proc. *122:* 269 (1963).

947 Wright, L.D. and Presberg, J.A.: Effect of certain compounds on solubility of cholesterol in coconut oil. Proc. Soc. exp. Biol. Med. *115:* 497–504 (1964).

948 Wruble, M.; Rundman, S.J., and Koning, J.H. (assigned to Upjohn Co.): Sitosterol oil-in-water emulsion for oral administration. US 3,085.939 (Apr.16, 1963).

949 Yamanouchi, M.: Extraction of phytosterols from soybean cake. Japan 2673 (Apr. 20, 1959).

950 Yamashita, K.: Progestational effect of androgenic steroids on the rabbit endometrium. J. Endocr. *30:* 271–272 (1964).

951 Yuasa, D.: Über Sterinresorption, gemessen am Pfortaderblut. Hoppe-Seyler's Z. physiol. Chem. *185:* 116–118 (1929).

952 Zayed, S.M.A.; Hassan, A., and Elghamry, M.I.: Estrogenic substances from Egyptian *Glycyrrhia glabra.* II. β-Sitosterol as an estrogenic principle. Zentbl. VetMed. *11:* ser. A, pp.476–482 (1964).

953 Zilletti, L.: Influenza del betasitosterolo sulla sintesi endogena del colesterolo. Boll. Soc. ital. Biol. sper. *422:* 933–936 (1966).

954 Zilletti, L.: Influence of β-sitosterol on cholesterol synthesis in the mouse liver *in vivo:* in Dumbovich, Proc. 4th Conf. Hungarian Soc. Ther. and Investig. Pharmacol., pp.405–407 (Akadémiai Kiadó, Budapest 1968).

955 Zilversmit, D.B.: A single blood sample method for dietary cholesterol absorption in the intact rat. Fed. Proc. *31:* 728 (1972).

956 Zilversmit, D.B.: A single blood sample dual method for the measurement of cholesterol absorption in rats. Proc. Soc. exp. Biol. Med. *140:* 862–865 (1972).

Subject Index